Intelligent Systems, Control and Automation: Science and Engineering

Volume 68

AF664821

For further volumes:
http://www.springer.com/series/6259

Matjaž Mihelj · Domen Novak
Samo Beguš

Virtual Reality Technology and Applications

Springer

Matjaž Mihelj
Domen Novak
Samo Beguš
Faculty of Electrical Engineering
University of Ljubljana
Ljubljana
Slovenia

ISSN 2213-8986 ISSN 2213-8994 (electronic)
ISBN 978-94-024-0036-6 ISBN 978-94-007-6910-6 (eBook)
DOI 10.1007/978-94-007-6910-6
Springer Dordrecht Heidelberg New York London

Printed on acid-free paper

Springer is part of Springer Science+Business Media (www.springer.com)

Preface

We began working on virtual reality in the first years of the twenty-first century, but that was not our first glimpse of it. As we grew up, we watched virtual reality through the eyes of laymen as it stepped out of science fiction and into everyday life. It has become a fascinating field that brings together engineers, programmers, designers, artists, psychologists, and others. These people collaborate to create something more than the sum of their parts, a virtual world made of zeros and ones that nonetheless feels real.

The magic of virtual worlds captivated us all, but what we desired most was to peer underneath the hood and see just how things worked. We thus became involved in the scientific and technical aspects of virtual reality: haptic interfaces, graphics design, psychological aspects, and others. We created this book for those who are also fascinated by the inner workings of this intriguing technology.

The book covers the individual elements of virtual reality, delving into their theory and implementation. It also describes how the elements are put together to create the virtual worlds we experience. Most of the knowledge contained within comes from our own experience in human–robot interaction, where virtual environments are used to entertain, motivate, and teach. Distilling the knowledge into text form has been an arduous process, and we leave it to readers to decide whether we were successful.

The text was originally aimed at engineers, researchers and graduate students with a solid foundation in mathematics. Our main motivation for writing it was that many existing virtual reality books do not have a sufficient focus on the technical, mathematical aspects that would be of interest to engineers. Nonetheless, the actual amount of mathematical content varies greatly from chapter to chapter. Readers with backgrounds other than engineering should be able to read and understand most chapters, though they may miss out on some of the mathematical details. Due to its origins, however, the book is focused less on psychological aspects and more on technical aspects—the hardware and software that makes virtual reality work.

Many people contributed either directly or indirectly to the creation of this book. Though they are too many to list, we would like to thank colleagues at the

University of Ljubljana and ETH Zurich, who travelled the path of research with us and helped us to discover virtual reality. We would also like to thank the diligent men and women at Springer who turned the book into reality. Cynthia Feenstra deserves special thanks for being in touch with us throughout the preparation process and putting up with occasionally missed deadlines. And as always, we would like to thank our families for supporting us day after day. Whoever you are, we hope you will enjoy reading this book.

Contents

Chapter 1
Introduction to Virtual Reality

Abstract The introductory chapter begins with the basic definitions of virtual reality, virtual presence and related concepts. It provides an overview of the history of virtual reality, from its origins in the 1950s to the present day. It also covers some of the most important applications, from flight simulators to biomedical uses. Finally, it briefly describes the main feedback loops used in virtual reality and the human biological systems used to interpret and act on information from the virtual world.

1.1 Definition of Virtual Reality

Virtual reality is a term that we've all heard many times. Movies such as the Matrix brought virtual reality out of science fiction and into the minds of the masses. Examples of virtual and augmented reality are also becoming more and more prevalent in real life, from military flight simulators to simple smartphone applications. But since everyone has their own impression of virtual reality is, let's first give the definition that we'll use throughout the book.

Virtual reality is composed of an interactive computer simulation, which senses the user's state and operation and replaces or augments sensory feedback information to one or more senses in a way that the user gets a sense of being immersed in the simulation (virtual environment). We can thus identify four basic elements of virtual reality: the virtual environment, virtual presence, sensory feedback (as a response to the user's actions) and interactivity [1].

1.1.1 Virtual Environment

A computer-generated *virtual environment* presents descriptions of objects within the simulation and the rules as well as relationships that govern these objects [2].

M. Mihelj et al., *Virtual Reality Technology and Applications*, Intelligent Systems, Control and Automation: Science and Engineering 68, DOI: 10.1007/978-94-007-6910-6_1,

Virtual reality is the observation of the virtual environment through a system that displays the objects and allows interaction, thus creating virtual presence.

Virtual environment is determined by its content (objects and characters). This content is displayed through various modalities (visual, aural and haptic), and perceived by the user through vision, hearing and touch.

Just like objects in the real world, objects in a virtual environment also have their properties such as shape, weight, color, texture, density and temperature. These properties can be observed using different senses. The color of an object, for example, is perceived only in the visual domain, while its texture can be perceived both in visual as well as haptic domains.

The content of the virtual environment can be grouped into categories. *Environment topology* describes the surface shape, areas and features. Actions in a virtual environment are usually limited to a small area within which the user can move. *Objects* are three-dimensional forms which occupy space in the virtual environment. They are entities that the user can observe and manipulate. *Intermediaries* are forms, which are controlled via interfaces, or avatars of users themselves. *User interface elements* represent parts of the interface that resides within the virtual environment. These include elements of virtual control such as virtual buttons, switches or sliders.

1.1.2 Virtual Presence

Virtual presence can be roughly divided into physical (sensory) and mental presence. It represents the feeling of actually 'being' in an environment; this can either be a completely psychological state or achieved via some physical medium. Physical virtual presence is the basic characteristic of virtual reality and represents the user's body physically entering the medium. Synthetic stimuli are created artificially and presented to the user, but it is not necessary to affect all senses or involve the entire human body. Mental virtual presence represents a state of 'trance': engagement, expectations, the feeling of being part of the virtual world. In addition to physical and mental presence, some authors also define telepresence, the feeling of virtual presence in a geographically distant location.

Virtual presence is very difficult to evoke with other media, as they do not offer actual sensory and physical immersion into the environment. The notion of *absence* has even been advanced as a concept analogous to presence, but evoked by other media [3]. Supporters of the concept claim that experiencing, for instance, the story of a novel requires a detachment from the environment in which the individual is reading a book. To some degree, information from the environment must be ignored so that the individual can be immersed in the contents of the novel—the reader must thus become absent from the surrounding environment. In virtual reality, the individual is present in an (admittedly virtual) environment, so he/she should also perceive it as real and respond to it as real.

1.1.2.1 Physical (sensory) Virtual Presence

Physical virtual presence defines virtual reality and simultaneously separates it from other media. It is achieved by presenting the virtual world to a user with a synthesis of stimuli to one or more senses in response to the user's position and actions. In general, a virtual reality system renders the virtual world through sight, sound and touch (haptics).

As the user moves, the visual, audio and haptic stimuli change as the virtual scene also moves. When moving toward an object, it becomes larger, louder, and can even be touched when at an appropriate distance. Turning the head shows the world to the left and right of the user. Touch allows objects to be manipulated.

Synthetic stimuli often drown out stimuli from the real world, thus decreasing mental presence in the real world. The degree to which real stimuli are replaced by synthetic ones and the number of 'tricked' senses affect the level of physical presence, which in turn affects the level of mental presence.

1.1.2.2 Mental Virtual Presence

The level of desired mental virtual presence depends on the goal applications of virtual reality. If the virtual experience is meant for entertainment, a high level of mental presence is needed. However, a high degree of mental immersion is often not necessary, possible or even desirable. The absence of mental virtual presence thus does not disqualify a medium from being virtual reality.

A user's mental virtual presence can have varying degrees of intensity: users can perceive a connection with the computer; users can ignore the real world and focus on interacting with the virtual world while still knowing the difference between real and virtual worlds; or users can even be so immersed in the virtual environment that they forget that it is virtual.

A realistic display that includes sight, sound and touch can significantly affect the level of mental virtual presence. A photorealistic image of the virtual environment is unnecessary and sometimes undesired, as even small errors in such an image distract the user from the experience. The same is true for other elements of realism such as three-dimensional views or echoes—while they are sometimes crucial for maintaining mental virtual presence, they may be distracting in other applications.

Virtual reality must ensure at least a minimum level of physical virtual presence. The definition of mental virtual presence assumes that users are so busy with events in the virtual environment that they stop doubting what they are experiencing. The level of mental virtual presence is affected by factors such as the virtual scenario, the quality of the display and graphical representation, and the number of senses stimulated by the virtual reality system. Another important factor is the delay between a user's action and the virtual environment's response. If the delay is too long (*being too long* depends on the display type—visual, aural or haptic), it can destroy the effect of mental immersion.

The perceived realism of individual objects and the entire virtual environment can be increased using *sensory transfer*. If an object looks realistic, we expect it to also act realistically. By emphasizing certain objects to the senses, it is possible to significantly increase the realism of the entire environment.

1.1.3 Sensory Feedback

Sensory feedback is a crucial component of virtual reality. The virtual reality system provides direct sensory feedback to users according to their physical location. Generally, most feedback is provided via visual information, though some environments only use haptic information. Of course, it is necessary to track the user's location in order to provide appropriate feedback. The system must thus have the ability to automatically measure the position and orientation of objects in the real environment.

1.1.4 Interactivity

If virtual reality is to be realistic, it must respond to the user's actions; in other words, it must be *interactive*. The ability of the user to affect computer-generated environments represents one form of interaction. Another possibility is to change the location and angle from which the user views the environment. A *multi-user environment* represents an extension to the interactivity and involves a large number of users simultaneously working in the same virtual environment or simulation. Thus, a multi-user environment must allow interaction between multiple users, but is not necessarily part of virtual reality.

When working with others in the same environment, it is necessary to follow their activities—pose, gestures, actions, gaze direction, speech. The word *avatar* (a Hindi word for the embodiment of a deity) is commonly used to describe a virtual object that represents a user or real object inside the virtual environment.

1.1.5 Perspective

The creator of virtual reality can use several options to change a user's perception of the virtual environment. One of them is the viewpoint from which the virtual world is seen. A first-person view involves looking at the environment through an avatar's eyes, a second-person view involves looking from the immediate vicinity of relevant activity, and a third-person view involves looking from an entirely independent location.

1.2 History of Virtual Reality

Human imagination dreamed of virtual reality for decades before the first actual implementations arrived. In 1931, Aldous Huxley's book *Brave New World* already introduced the concept of *feelies*—movies that involve touch in addition to sight and sound. In 1935, Stanley Weinbaum went even farther and presented a detailed idea of virtual reality in his book *Pygmalion's spectacles*:

"But listen—a movie that gives one sight and sound. Suppose now I add taste, smell, even touch, if your interest is taken by the story. Suppose I make it so that you are in the story, you speak to the shadows, and the shadows reply, and instead of being on a screen, the story is all about you, and you are in it. Would that be to make real a dream?"

This idea remained on paper for about 20 years until Morton Heilig, considered by some to be the father of virtual reality, began designing a practical implementation. In 1957, he developed and patented the *Sensorama*, a machine that offered a virtual bicycle riding experience. The user sat in the machine with a three-dimensional city display, heard the sounds of the city, felt the wind and the vibrations of the seat, and could even smell certain scents. The Sensorama was the first step in the real development of virtual reality, but was never commercially successful.

The next device worth mentioning was the first head-mounted display, the Philco HMD. It did not provide a window into a virtual environment, but showed a video from a real, distant location. It can thus be considered the first example of telepresence, an application of virtual reality that is still popular today. In 1968, Ivan Sutherland then developed a head-mounted display connected to a virtual environment [4]. Sutherland's paper *The Ultimate Display*, written in 1965, predicted the rise of the fantastic worlds seen in today's computer games: "There is no reason why the objects displayed by a computer have to follow the ordinary rules of physical reality with which we are familiar." He also had a vision of the ultimate stage of virtual reality development:

"The ultimate display would, of course, be a room within which the computer can control the existence of matter. A chair displayed in such a room would be good enough to sit in. Handcuffs displayed in such a room would be confining, and a bullet displayed in such a room would be fatal. With appropriate programming, such a display could literally be the Wonderland into which Alice walked."

Sutherland's display, called the *Sword of Damocles*, consisted of glasses with two small screens (one for each eye) that together gave an illusion of three-dimensional vision. It displayed a virtual environment consisting of rooms represented with simple wire models. By moving the head, the user could also change the view of the environment, which required a complex motion tracking system attached to the ceiling. Since the screens were partially transparent, the user could see both the real and virtual worlds simultaneously. The Sword of Damocles can thus also be considered the first example of *augmented reality*: synthetic stimuli superimposed on stimuli from the real environment.

Sensorama and the Sword of Damocles allowed a user to experience virtual environments using different senses, but did not allow any interaction with these environments. The first environments that reacted to the user's actions were developed around 1970 by Myron Krueger. Using various sensors (from videocameras to pressure sensors in the floor), the virtual reality system could recognize users' activities and move objects in the virtual environment accordingly. Virtual objects thus acted like real ones. Since multiple users could interact with the virtual environment simultaneously, this was also the first example of *multi-user environments*. Krueger's most famous creation was the *Videoplace* environment, which included artistic activities such as drawing on virtual objects. Krueger also coined the term *artificial reality*, which describes the recognition of the user's activities and the generation of feedback that reinforces the illusion of the activities taking place in a virtual environment.

Virtual environments that could react to the user's actions required new motion recognition methods adapted for virtual reality. The late 1960s and early 1970s thus saw the development of the *Grope I–III* systems for virtual molecule display. They allowed the user to move molecules on the screen using a special haptic interface as well as feel the forces between the molecules using force feedback. Grope I–III were an upgrade of Krueger's system: instead of recognizing motions using videocameras, they let the user affect objects directly by touching them. The next step in touch recognition was the *Sayre Glove*, which used built-in sensors to detect finger motions and thus represented a simple and cheap human motion recognition method. Its younger sister, the *VPL Dataglove*, appeared on a 1987 cover of Scientific American, entering public consciousness as the first commercially available motion recognition glove. VPL, the company that developed it, later also developed the first commercial virtual reality system, *Reality built for two*. The company's founder Jaron Lanier also significantly contributed to public awareness of virtual reality and popularization of the term virtual reality itself.

Of course, it was not only the scientists that brought the concept of virtual reality to the masses. Science fiction books, TV series and movies all did their part. In 1982, William Gibson first used the term *cyberspace* in his stories Burning Chrome and Neuromancer. The same year, Hollywood first brought virtual reality to the silver screen with the movie Tron. In 1987, the TV series Star Trek: The Next Generation presented the *holodeck*: a room that used holographic images and the creation of new matter to conjure up virtual environments that users could actively participate in. One episode of the series even centered on a man who became so obsessed with virtual reality that he neglected the real world; a shadowy dream of modern massively multiplayer online games? The 1990s then delivered an abundance of movies about virtual reality. Perhaps the most famous one was the Matrix, which imagined a virtual world that encompassed the majority of humanity and was so realistic that most users did not even know it was only virtual.

Technological development thus fell far behind human imagination in the 1990s, but major progress was made nonetheless. Perhaps the most famous virtual reality product of the nineties was the *CAVE* (an acronym that stands for *Cave Automatic Virtual Environment*): a room whose walls consist of screens displaying a virtual environment. Users can thus really see themselves inside the virtual environment. Special glasses can also give an illusion of depth—objects look like they step out of the screen and float in the air. Electromagnetic sensors built into the walls allow the measurement of motions while a surround sound system allows three-dimensional sound. The CAVE has become quite popular and is used by universities worldwide. Due to the high price, home versions are not yet available, but the same technologies is used in amusement parks and movie theaters as the *3D cinema*. CAVE may be the first simple implementation of the *holodeck*, but it will still be a long way to Star Trek's vision.

1.3 Applications of Virtual Reality

Although virtual reality has not yet achieved the visions of science fiction, it is already successfully used in many applications. Let's briefly examine some of them, as a detailed overview of all applications would take up far too much space.

1.3.1 Flight and Driving Simulators

Flight simulators may be the best-known practical application of virtual reality. They allow pilots to practice flying in a safe, controlled environment where mistakes cannot lead to injury or equipment damage. The simplest simulators run on a personal computer. Their virtual environment contains the entire physical model of the plane and a simpler model of the landscape over which the plane can fly. More complex simulators use similar virtual environments, but include a realistic imitation of the cockpit complete with visual displays, sounds and mechanisms to move the entire cockpit (when simulating e.g. turbulence). Such virtual environments allow training in varied situations, from routine flight to serious equipment failures that could endanger passengers' lives. Since the desired situation can be chosen at will, training in a flight simulator can also be more effective per unit of time than real training. Flight simulators first appeared in the 1950s and now represent a completely established technology regularly used by military and civilian pilots all over the world.

Driving simulators were developed for a similar purpose: they allow safe driving lessons in different conditions (rain, ice, congestions) or tests of new cars. In a virtual environment, it is possible to change any of the car's features (both aesthetic and functional) and then observe how real drivers react to the changes. Simulation thus also allows designed cars to be tested before actually building a prototype.

1.3.2 Surgery Simulators

A flight simulator is a typical example of virtual reality since it allows difficult tasks to be practiced in a controlled environment where different actions and conditions can be tried without any threat to people or machinery. Surgery similarly represents a difficult situation where a single error can lead to the patient's death. Following in the footsteps of flight simulators, surgery simulators provide virtual environments where a surgeon can use realistic haptic interfaces (which look and feel like actual surgical instruments) to practice surgical procedures on different patients [5]. These virtual patients are also not necessarily just 'general', made-up people. Information obtained with modern medical equipment (e.g. computed tomography) can be used to create a three-dimensional model of a patient that is scheduled for actual surgery. Before the real surgery, surgeons can 'practice' on a virtual patient with very similar characteristics to the real one. Surgery simulators have become especially widespread with the creation of surgical robots, which allow the entire surgery to be conducted via haptic interface and screen. With surgical robots, experience from virtual surgery is even more directly transferrable into reality.

1.3.3 Design and Visualization

As mentioned in the driving simulator subsection, virtual reality can be used to design and test different machines and objects. Since virtual reality is often expensive, it is most frequently used to design objects that are either very expensive (e.g. power plants, rockets) or manufactured in large quantities (e.g. cars). Such virtual environments are extremely complex since they need to combine a good visual display with a detailed physical model that includes all the factors that could affect the tested object.

Designing objects in virtual reality does not have to be limited to testing concepts that could later be transferred to the real world. The process can also go the other way: objects that exist in the real world can be transferred to a virtual environment. One example are imitations of famous buildings created in virtual environments [6]. The user can walk through a virtual historical building, play with the items in it and learn historical facts without ever visiting the building in person. The virtual environment can even include virtual humans from the historical era of the building, allowing the user to interact with them and learn additional information.

1.3.4 Telepresence and Teleoperation

Telepresence is the use of virtual reality to virtually place the user in some distant real place. The user has the ability to directly interact (often via a computerized medium) with this place, regardless of where it is. *Teleoperation* differs from telepresence in that the operator does not interact with the distant place directly, but uses some device located in the distant place.

Teleoperation is primarily used to operate robots at a distance. The best-known examples are mobile robots used to explore hazardous environments: the Moon, Mars or simply dangerous places on Earth. The robot has a camera attached to its frame and is controlled by an operator located in a safe place that can even be millions of kilometers away. Similar technology is used to control unmanned military drones. These robots and planes usually have a certain level of autonomy, allowing them to react to simple problems while having the operator carry out more difficult tasks. Another example was already mentioned in surgery: minimally invasive surgical robots guided by a surgeon via haptic interface and cameras. The surgeon can be in the same room, in an adjoining room or even far away. The first *transatlantic* surgical procedure was performed in 2011—the surgeon was in New York while the robot and patient were in Strasbourg.

1.3.5 Augmented Reality

Augmented reality is an upgrade of virtual reality where synthetic stimuli (computer-generated visual, audio or haptic information) are superimposed onto real-world stimuli. These integrated stimuli affect users directly or via a display, allowing them to perceive otherwise invisible information. An example from surgery is the artificial displaying of information from the interior of the body (e.g. skeletal structure) on the appropriate part of the skin surface. Besides surgery, augmented reality is also used e.g. in industrial and military devices to display important information on screens. In the last few years, it has also spread to fields such as tourism and advertising [7]. Augmented reality is thus often successfully implemented in mobile phones that can sense the user's position and provide information such as an image of a nearby street (real stimulus) with all stores or landmarks marked (computer-generated visual stimulus).

1.3.6 Motor Rehabilitation

After injuries such as stroke or spinal cord injury, human limbs are severely weakened due to a damaged nervous system. Intensive exercise can help the patient partially or completely regain lost motor abilities, but the rehabilitation process is lengthy and difficult. Patients need to obtain detailed feedback both during and after exercise in order to improve their motions, and they must be highly motivated for exercise. Virtual reality has been suggested as a possible solution to these problems. Interesting and varied virtual environments can increase motivation since the exercises do not become monotonous. At the same time, measuring all the variables of the virtual environment allows a large amount of information about patient movements and general performance to be obtained. Virtual reality can also be combined with special rehabilitation robots that actively help the patient exercise. Several studies have shown that patients can relearn motions in virtual reality and that the knowledge can be successfully transferred to the real world. However, it has not yet been proven whether rehabilitation in virtual reality is more effective than classic rehabilitation methods [8].

1.3.7 Psychotherapy

Virtual reality can evoke virtual presence: a feeling of being present and involved in the virtual environment. Users thus react to virtual objects and creatures as they would to real ones, a feature often used by psychologists. In fact, the most popular therapeutic application of virtual reality is the treatment of phobias and traumas by exposing a individual to the object, creature or situation that they are afraid of. Psychology discovered long ago that people will never overcome their fears if they avoid stressful situations—they must face their fears in order to overcome them. Since exposing an individual to an actual stressful situation can be expensive, dangerous, impractical or even impossible (e.g. in the case of post-traumatic stress disorder evoked by wars), virtual reality can act as an effective alternative. Exposure therapy in virtual reality is completely controlled, inexpensive and can be performed at a therapeutic institution. Virtual reality has thus been used to successfully cure fear of heights, spiders, flying, open spaces and public speaking. Additionally, virtual environments with many positive stimuli can be used to treat other psychological disorders such as impotence or low self-esteem caused by excess weight [9].

1.4 Virtual Reality System

Virtual reality relies on the use of a feedback loop. Figure 1.1 shows the feedback loop, which allows interaction with the virtual reality system through user's physical actions and detection of user's psychophysiological state. In a *fast* feedback loop the user directly interacts with the virtual reality system through motion. In a *slow* feedback loop related to affective computing, the psychophysiological state of the user can be assessed through measurements and analysis of physiological signals and the virtual environment can be adapted to engage and motivate the user.

The virtual reality system enables exchange of information with the virtual environment. Information is exchanged through the interface to the virtual world. The *user interface* is the gateway between the user and the virtual environment. Ideally, the gateway would allow transparent communication and transfer of information between the user and the virtual environment.

The user interface defines how the user communicates with the virtual world and how the virtual world manifests in a perceptible way. Figure 1.2 shows the relationships between the user interface, methods of creating the virtual world and aspects of the user's personality. All these elements affect the virtual reality experience as well as the physical and mental presence.

Figure 1.3 shows the flow of information in a typical virtual reality system. The virtual world is projected into a representation that is rendered and shown to the user via displays. The process takes the user's motions into account, thus enabling virtual presence by appropriately adjusting the displayed information. The user can affect the virtual world via the interface inputs. In augmented reality, the displayed virtual environment is superimposed onto the perceived real environment.

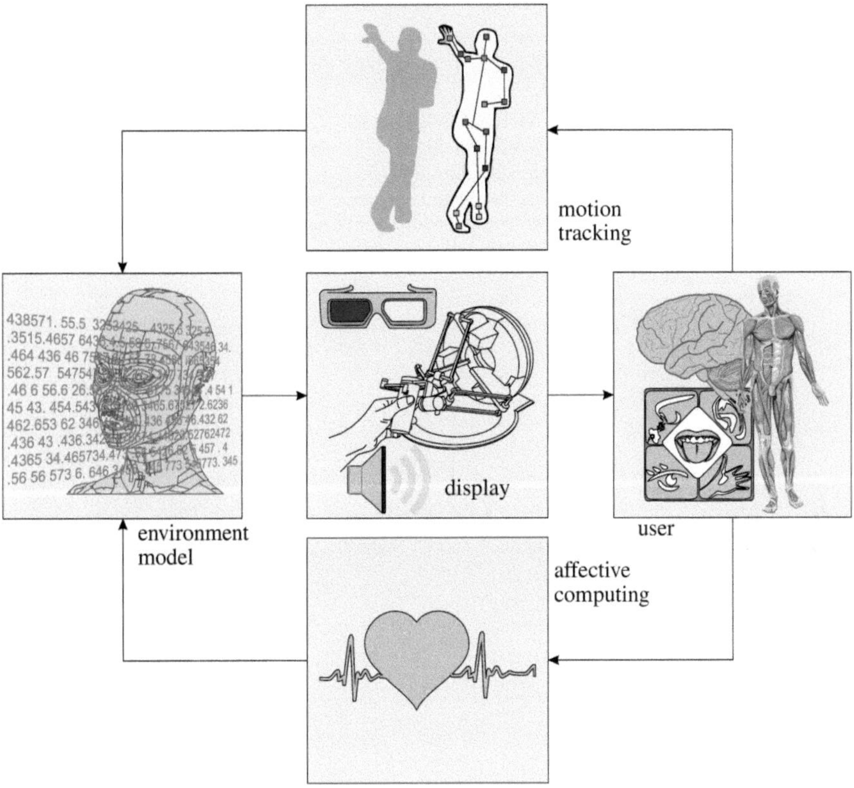

Fig. 1.1 The feedback loop is a crucial element of a virtual reality system. The system must react to the user's actions, and can optionally even estimate the user's psychological state in order to better adapt to the situation

1.4.1 Representation, Rendering and Displaying of the Virtual World

Rendering is the process of creating sensory images of the virtual world. They must be refreshed rapidly enough to give the user an impression of continuous flow (real-time rendering). The creation of sensory images consists of two steps. First, it is necessary to decide how the virtual world should look, sound and feel. This is called the representation of the virtual world. Secondly, the representation must be displayed using appropriate hardware and software.

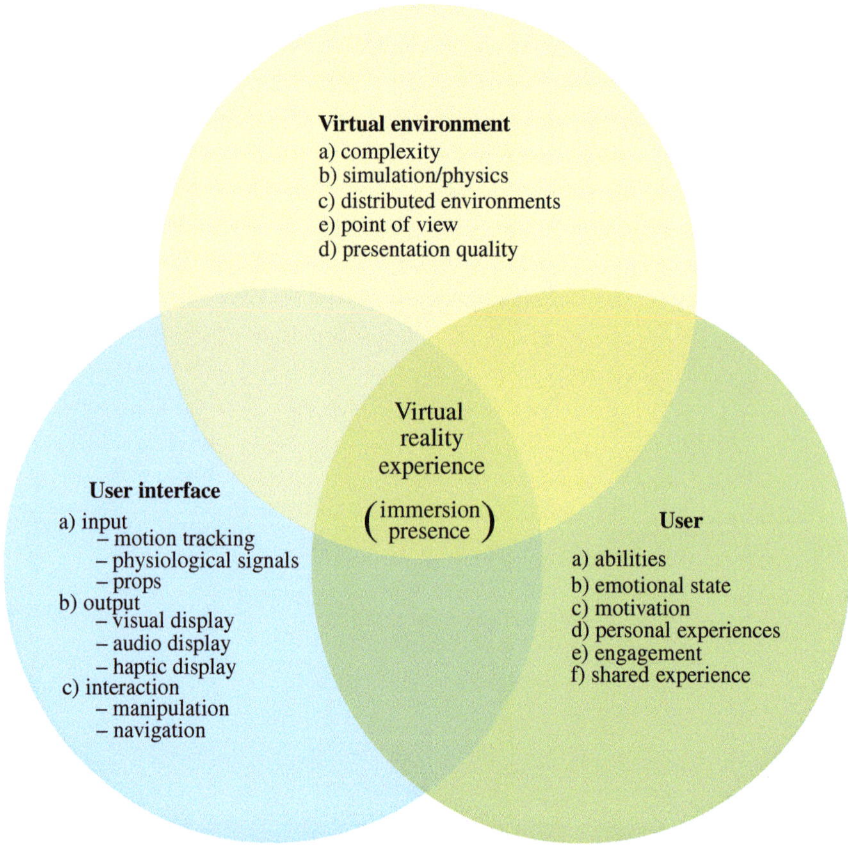

Fig. 1.2 Virtual reality requires the integration of multiple factors—user interface, elements of the virtual world and the user's experiences. Interaction between these factors defines the experience of virtual reality

1.4.1.1 Representation of the Virtual World

If we wish to create virtual reality, we must decide how to represent thoughts, ideas and information in visual, audio and haptic forms. This decision significantly affects the effectiveness of virtual reality.

Communication via a certain medium thus demands mutual presentation of ideas and the understanding of these presentations. Ideas, concepts or physical characteristics can be presented in different ways, though some are more appropriate than others.

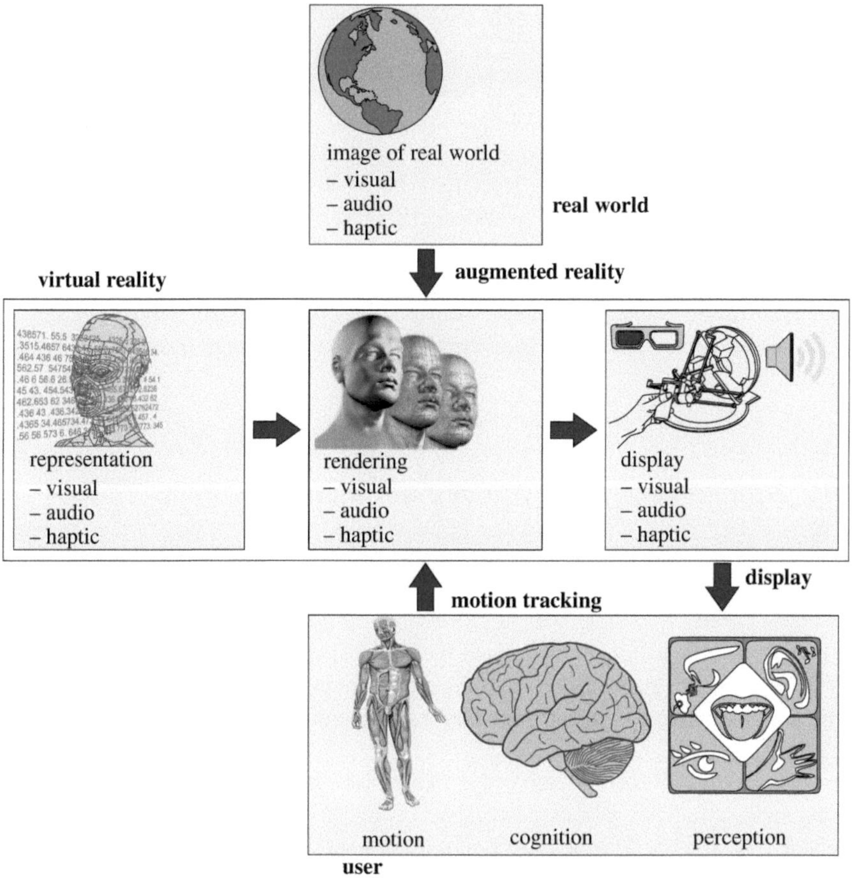

Fig. 1.3 Diagram of information flow in a virtual reality system

1.4.1.2 Rendering of the Virtual World

Rendering generates visual, audio and haptic signals to be displayed with appropriate equipment. Hardware and software allow computer-generated representations of the virtual world to be transformed into signals that are then displayed in a way noticeable to the human senses. Since each sense has different rendering requirements, different hardware and software are thus also used for each sensory channel. Though the ultimate goal is to create an unified virtual world for all the senses, the implementation details differ greatly and will thus be covered separately for visual, audio and haptic renderings.

1.4.1.3 Display Technologies

The experience of virtual reality is based on the user's perception of the virtual world, and physical perception of the virtual world is based on computer displays. The term *display* will be used throughout the book to refer to any method of presenting information to any human sense. The human body has at least five senses that provide information about the external world to the brain. Three of these senses—sight, hearing and touch—are the most frequently used to transmit synthetic stimuli in virtual reality. The virtual reality system 'tricks' the senses by displaying computer-generated stimuli that replace stimuli from the real world. In general, the more senses are provided with synthetically generated stimuli, the better the experience of virtual reality.

1.4.2 Human Perception, Motor and Cognitive Systems

From the perspective of interaction with a virtual environment, a human being can be divided into three main systems:

- *perception* allows information about the environment to be obtained;
- *motor abilities* (musculoskeletal system) allow movement through the environment, manipulation through touch, and positioning of sensory organs for better perception;
- *cognitive abilities* (central nervous system) allow the analysis of information from the environment and action planning according to current task goals.

1.4.2.1 Human Perception

Humans perceive the environment around them via multiple sensory channels that allow electromagnetic (sight), chemical (taste, smell), mechanical (hearing, touch, orientation) and heat stimuli to be detected. Many such stimuli can be artificially reproduced in virtual reality.

Biological converters that receive signals from the environment or the interior of the body and transmit them to the central nervous system are called receptors. In general, each receptor senses only one type of energy or stimulus. Receptor structure is thus also very varied and adapted to receiving specific stimuli. Nonetheless, despite this great variety, most receptors can be divided into the three functional units shown in Fig. 1.4. The input signal is a stimulus that always appears in the form of energy: electromagnetic, mechanical, chemical or heat. The stimulus affects the filter part of the receptor, which does not change the form of the energy but does amplify or suppress certain parameters. For instance, the ear amplifies certain frequencies of sound, skin acts as a mechanical filter, and the eyes use the lens to focus light onto the

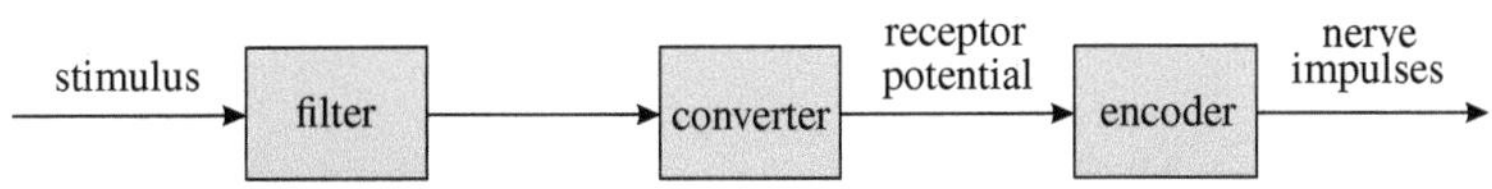

Fig. 1.4 Functional units of a receptor

retina. The converter changes the modified stimulus to a receptor membrane potential while the encoder finally converts the signal to a sequence of action potentials.

Most receptors decrease the output signal if the input does not change for a certain amount of time. This is called adaptation. The receptor's response usually has two components: the first is proportional to the intensity of the stimulus while the second is proportional to the speed with which the stimulus intensity changes. If $S(t)$ is the stimulus intensity and $R(t)$ is the response, then

$$R(t) = \alpha S(t) + \beta \dot{S}(t), \tag{1.1}$$

where α and β are constants or functions that represent adaptation.

Several stimuli can be better related to the response through the *Weber-Fechner* equation:

$$R = K \log \frac{S}{S_0}, \tag{1.2}$$

where K is a constant and S_0 is the threshold value of the stimulus.

1.4.2.2 Human Motor System

Humans use their motor abilities to interact with the virtual world. This interaction can be roughly divided into navigation within the virtual environment, manipulation of objects in the environment, and interaction with other users of the virtual reality. The motor subsystem also has an important role in connection to haptic interfaces. The user directly interacts with the haptic interface and thus affects the stability of haptic interaction. It is thus necessary to be familiar with the human motor system in order to create a stable haptic interface.

1.4.2.3 Human Cognitive System

The human cognitive system is used to make decisions about how to interact with the virtual environment. In addition to the rational part of cognition, it is also crucial to take emotions into account since they have an important effect on behavior in the virtual world.

References

1. Sherman WR, Craig AB (2003) Understanding virtual reality. Morgen Kaufman Publishers
2. Stanney K (2001) Handbook of virtual environments. Lawrence Earlbaum, Inc
3. Waterworth JA, Waterworth EL (2003) The meaning of presence. Presence-Connect 3(2)
4. Sutherland IE (1965) The ultimate display. In: Proceedings of the IFIP congress, pp 506–508
5. Gallagher AG, Ritter EM, Champion H, Higgins G, Fried MP, Moses G, Smith D, Satava RM (2005) Virtual reality simulation for the operating room proficiency-based training as a paradigm shift in surgical skills training. Ann Surgery 241(2):364–372
6. Anderson EF, McLoughlin L, Liarokapis F, Peters C, Petridis P, de Freitas S (2010) Developing serious games for cultural heritage: a state-of-the-art review. Virtual Reality 14:255–275
7. Yu D, Jin JS, Luo S, Lai W, Huang Q (2010) Visual information communication, chap. A useful visualization technique: a literature review for augmented reality and its application, limitation and future direction, Springer, pp 311–337
8. Holden MK (2005) Virtual environments for motor rehabilitation: review. CyberPsychology Behav 8(3):187–211
9. Riva G (2005) Virtual reality in psychotherapy: review. CyberPsychology Behav 8(3):220–230

Chapter 2
Degrees of Freedom, Pose, Displacement and Perspective

Abstract A virtual environment is composed of entities such as objects, avatars and sound sources. The number of independent axes of translation and rotation along and around which an entity can move defines the number of degrees of freedom of the particular entity. The pose (position and orientation) and displacement (translation and rotation) of an entity can be defined with the use of homogenous transformation matrices. Rotations and orientations can also be expressed in terms of Euler angles or quaternions. Homogenous transformation matrices also make it possible to define perspective transformations. In this chapter the basic mathematical tools in the study of rigid body kinematics will be introduced.

A virtual environment is composed of entities such as objects, avatars and sound sources. The positions and orientations of these entities in space are determined by task requirements and interaction with the user of a virtual reality system. On the other hand, the user is positioned in the real world, so his/her pose is also determined relative to a reference coordinate frame (for example, a coordinate frame of a position tracking device that tracks the movement of the user). Entities of the real world as well as entities of a virtual environment can move within the boundaries defined by the equipment and the specifications of the task.

The number of independent axes of translation and rotation along and around which an entity can move defines the number of degrees of freedom of the particular entity. The pose and displacement of an entity can be defined with the use of homogenous transformation matrices. These matrices also make it possible to define perspective transformations, which are relevant for presentation of a virtual environment from different perspectives (for example, from the user's perspective). The chapter provides a basic course on homogenous transformations and quaternions. More information can be found in [1, 2].

M. Mihelj et al., *Virtual Reality Technology and Applications*,
Intelligent Systems, Control and Automation: Science and Engineering 68,
DOI: 10.1007/978-94-007-6910-6_2,

2.1 Degree of Freedom

To begin with, we will introduce the degree of freedom in the case of an infinitesimal mass particle. In this case, the number of degrees of freedom is defined as the number of independent coordinates (not including time) which are necessary for the complete description of the position of a mass particle.

A particle moving along a line (infinitesimally small ball on a wire) is a system with one degree of freedom. A pendulum with a rigid segment swinging in a plane is also a system with one degree of freedom (Fig. 2.1). In the first example, the position of the particle can be described with the distance, while in the second case it is described with the angle of rotation.

A mass particle moving on a plane has two degrees of freedom (Fig. 2.2). Its position can be described with two cartesian coordinates x and y. A double pendulum with rigid segments swinging in a plane is also a system with two degrees of freedom. The position of the mass particle is described by two angles. A mass particle in space has three degrees of freedom. Its position is usually expressed by three Cartesian coordinates x, y and z. An example of a simple mechanical system with three degrees of freedom is a double pendulum where one segment is represented by an elastic spring and the other by a rigid rod. In this case, the pendulum also swings in a plane.

Next, we will consider degrees of freedom of a rigid body. The simplest rigid body consists of three mass particles (Fig. 2.3). We already know that a single mass particle has three degrees of freedom described by three rectangular displacements along a line called translations (T). We add another mass particle to the first one in such a way that there is constant distance between them. The second particle is

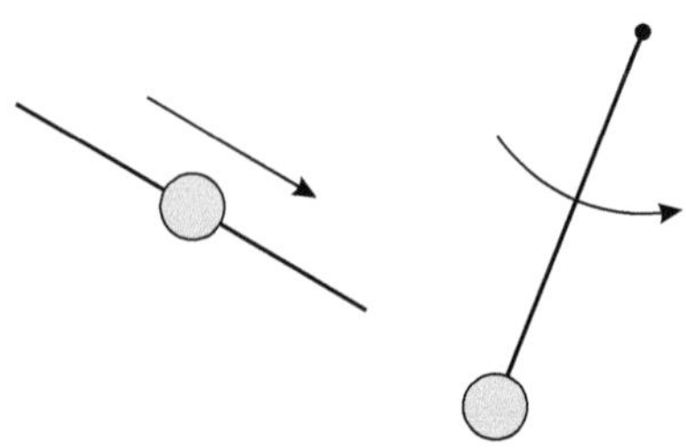

Fig. 2.1 Two examples of systems with one degree of freedom: mass particle on a wire (*left*) and rigid pendulum in a plane (*right*)

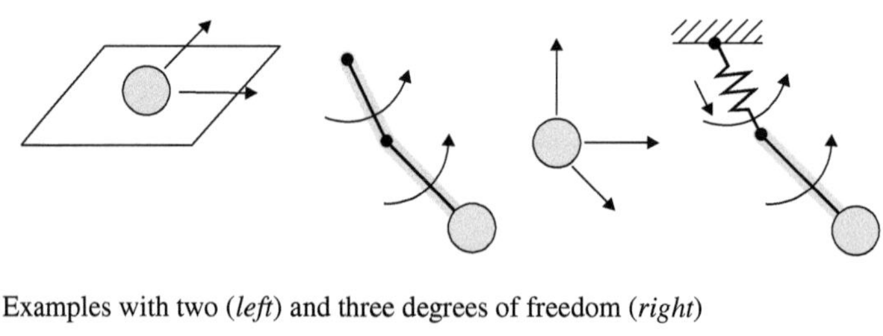

Fig. 2.2 Examples with two (*left*) and three degrees of freedom (*right*)

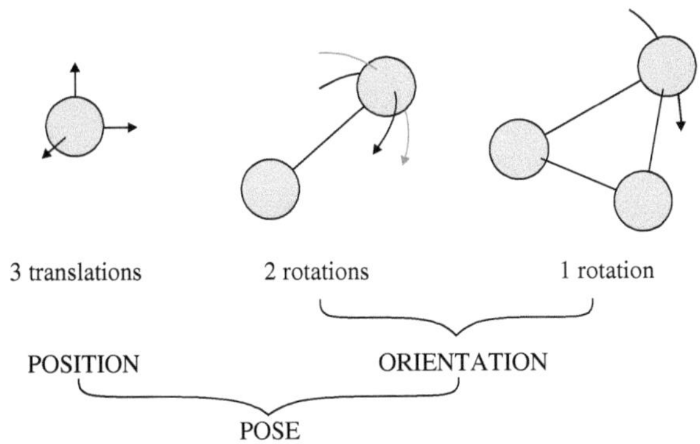

Fig. 2.3 Degrees of freedom of a rigid body

restricted to move on the surface of a sphere surrounding the first particle. Its position on the sphere can be described by two circles reminiscent of meridians and latitudes on a globe. The displacement along a circular line is called rotation (R). The third mass particle is added in such a way that the distances with respect to the first two particles are kept constant. In this way the third particle may move along a circle, a kind of equator, around the axis determined by the first two particles. A rigid body therefore has 6 degrees of freedom: 3 translations and 3 rotations. The first three degrees of freedom describe the position of the body while the other three degrees of freedom determine its orientation. The term pose is used to include both position and orientation.

In the following sections we will introduce a unified mathematical description of translational and rotational displacements.

2.2 Translational Transformation

The translational displacement $\mathbf{d}$, given by the vector

$$\mathbf{d} = a\mathbf{i} + b\mathbf{j} + c\mathbf{k}, \tag{2.1}$$

can also be described also by the following homogenous transformation matrix $\mathbf{H}$

$$\mathbf{H} = Trans(a, b, c) = \begin{bmatrix} 1 & 0 & 0 & a \\ 0 & 1 & 0 & b \\ 0 & 0 & 1 & c \\ 0 & 0 & 0 & 1 \end{bmatrix}. \tag{2.2}$$

When using homogenous transformation matrices, an arbitrary vector has the following 4×1 form

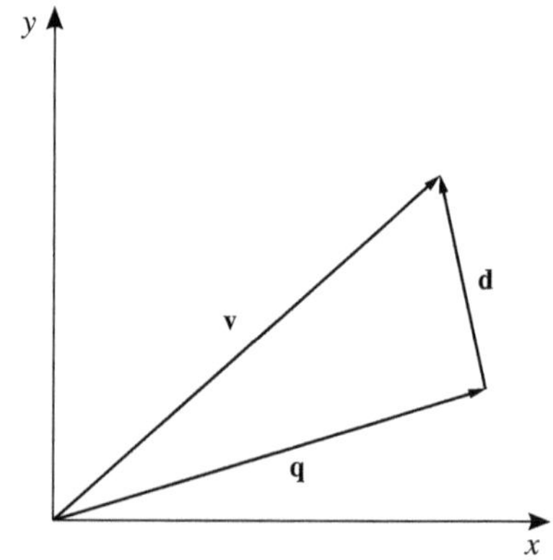

Fig. 2.4 Translational transformation

$$\mathbf{q} = \begin{bmatrix} x \\ y \\ z \\ 1 \end{bmatrix} = \begin{bmatrix} x & y & z & 1 \end{bmatrix}^T. \tag{2.3}$$

A translational displacement of vector **q** for a distance **d** is obtained by multiplying the vector **q** with the matrix **H**

$$\mathbf{v} = \begin{bmatrix} 1 & 0 & 0 & a \\ 0 & 1 & 0 & b \\ 0 & 0 & 1 & c \\ 0 & 0 & 0 & 1 \end{bmatrix} \begin{bmatrix} x \\ y \\ z \\ 1 \end{bmatrix} = \begin{bmatrix} x+a \\ y+b \\ z+c \\ 1 \end{bmatrix}. \tag{2.4}$$

The translation, which is presented by multiplication with a homogenous matrix, is equivalent to the sum of vectors **q** and **d** (Fig. 2.4)

$$\mathbf{v} = \mathbf{q} + \mathbf{d} = (x\mathbf{i} + y\mathbf{j} + z\mathbf{k}) + (a\mathbf{i} + b\mathbf{j} + c\mathbf{k}) = (x+a)\mathbf{i} + (y+b)\mathbf{j} + (z+c)\mathbf{k}. \tag{2.5}$$

In a simple example, the vector $2\mathbf{i} + 3\mathbf{j} + 2\mathbf{k}$ is translationally displaced by the distance $4\mathbf{i} - 3\mathbf{j} + 7\mathbf{k}$

$$\mathbf{v} = \begin{bmatrix} 1 & 0 & 0 & 4 \\ 0 & 1 & 0 & -3 \\ 0 & 0 & 1 & 7 \\ 0 & 0 & 0 & 1 \end{bmatrix} \begin{bmatrix} 2 \\ 3 \\ 2 \\ 1 \end{bmatrix} = \begin{bmatrix} 6 \\ 0 \\ 9 \\ 1 \end{bmatrix}.$$

The same result is obtained by adding the two vectors.

2.3 Rotational Transformation

Rotational displacements will be described in a right-handed Cartesian coordinate frame where the rotations around the three axes, as shown in Fig. 2.5, are considered as positive. Positive rotations around the selected axis are counter-clockwise when looking from the positive end of the axis towards the origin of the frame O. The positive rotation can be described also by the so-called right-hand rule where the thumb is directed along the axis towards its positive end while the fingers show the positive direction of the rotational displacement.

Let us first take a closer look at the rotation around the x axis. The coordinate frame x', y', z' shown in Fig. 2.6 was obtained by rotating the reference frame x, y, z in the positive direction around the x axis for the angle α. The axes x and x' are collinear.

The rotational displacement is also described by a homogenous transformation matrix. The first three rows of the transformation matrix correspond to the x, y and z axes of the reference frame while the first three columns refer to the x', y' and z' axes of the rotated frame. The upper left nine elements of the matrix $\mathbf{H}$ represent the

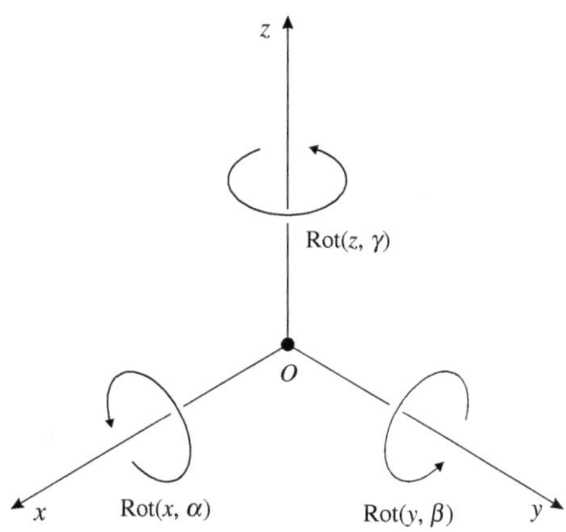

Fig. 2.5 Right-hand rectangular frame with positive rotations

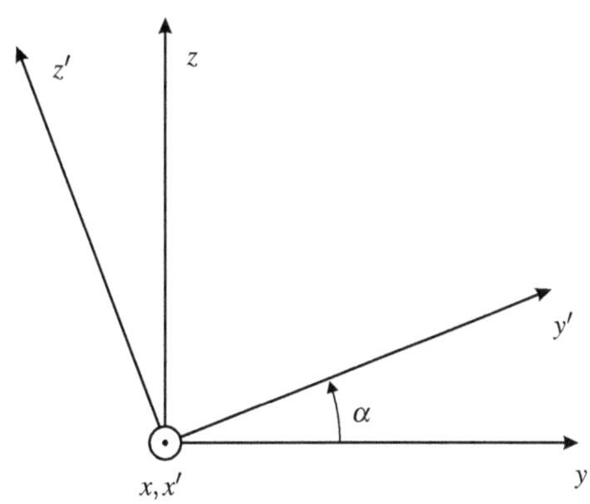

Fig. 2.6 Rotation around x axis

3×3 rotation matrix. The elements of the rotation matrix are cosines of the angles between the axes given by the corresponding column and row

$$Rot(x, \alpha) = \begin{array}{c} \\ \\ \\ \\ \end{array}\begin{matrix} & x' & y' & z' & \\ \end{matrix}$$

$$Rot(x, \alpha) = \left[\begin{array}{cccc} \cos 0° & \cos 90° & \cos 90° & 0 \\ \cos 90° & \cos \alpha & \cos(90° + \alpha) & 0 \\ \cos 90° & \cos(90° - \alpha) & \cos \alpha & 0 \\ 0 & 0 & 0 & 1 \end{array}\right] \begin{array}{c} x \\ y \\ z \\ \\ \end{array} \quad (\text{columns } x', y', z')$$

$$= \left[\begin{array}{cccc} 1 & 0 & 0 & 0 \\ 0 & \cos \alpha & -\sin \alpha & 0 \\ 0 & \sin \alpha & \cos \alpha & 0 \\ 0 & 0 & 0 & 1 \end{array}\right]. \tag{2.6}$$

The angle between the x' and the x axes is $0°$. We therefore have $\cos 0°$ in the intersection of the x' column and the x row. The angle between the x' and the y axes is $90°$. We put $\cos 90°$ in the corresponding intersection. The angle between the y' and the y axes is α. The corresponding matrix element is $\cos \alpha$.

Rotation matrices for rotations around the y axis can be written similarly

$$Rot(y, \beta) = \left[\begin{array}{cccc} \cos \beta & 0 & \sin \beta & 0 \\ 0 & 1 & 0 & 0 \\ -\sin \beta & 0 & \cos \beta & 0 \\ 0 & 0 & 0 & 1 \end{array}\right] \tag{2.7}$$

and z axis

$$Rot(z, \gamma) = \left[\begin{array}{cccc} \cos \gamma & -\sin \gamma & 0 & 0 \\ \sin \gamma & \cos \gamma & 0 & 0 \\ 0 & 0 & 1 & 0 \\ 0 & 0 & 0 & 1 \end{array}\right]. \tag{2.8}$$

In a simple numerical example, we wish to determine the vector $\mathbf{w}$ which is obtained by rotating the vector $\mathbf{u} = 3\mathbf{i} + 2\mathbf{j} + 0\mathbf{k}$ for $90°$ in the counter-clockwise i.e. positive direction around the z axis. As $\cos 90° = 0$ and $\sin 90° = 1$, it is not difficult to determine the matrix describing $Rot(z, 90°)$ and multiplying it by the vector $\mathbf{u}$

$$\mathbf{w} = \left[\begin{array}{cccc} 0 & -1 & 0 & 0 \\ 1 & 0 & 0 & 0 \\ 0 & 0 & 1 & 0 \\ 0 & 0 & 0 & 1 \end{array}\right] \left[\begin{array}{c} 3 \\ 2 \\ 0 \\ 1 \end{array}\right] = \left[\begin{array}{c} -2 \\ 3 \\ 0 \\ 1 \end{array}\right].$$

The graphical presentation of rotating the vector $\mathbf{u}$ around the z axis is shown in Fig. 2.7.

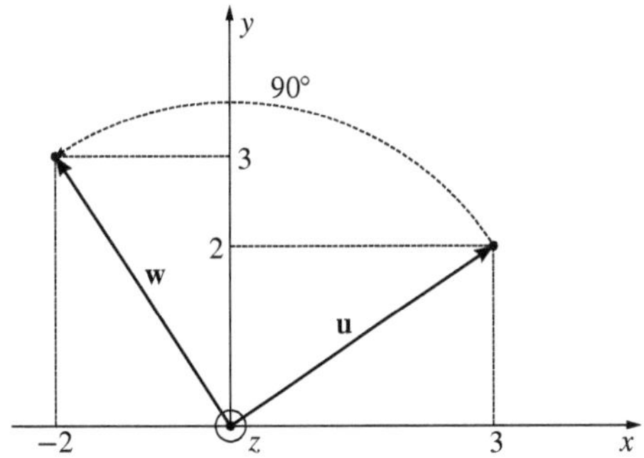

Fig. 2.7 Example of rotational transformation

2.4 Pose and Displacement

In the previous section, we learned how a point is translated or rotated around the axes of the cartesian frame. We are next interested in displacements of objects. We can always attach a coordinate frame to a rigid object under consideration. In this section we shall deal with the pose and the displacement of rectangular frames. We shall see that a homogenous transformation matrix

$$\mathbf{H} = \begin{bmatrix} \mathbf{R} & \mathbf{p} \\ \mathbf{0} & 1 \end{bmatrix} \tag{2.9}$$

describes either the pose of a frame with respect to a reference frame or it represents the displacement of a frame into a new pose. In the first case the upper left 3×3 matrix $\mathbf{R}$ represents the orientation of the object, while the right-hand 3×1 column $\mathbf{p}$ describes its position (e.g. the position of its center of mass). In the case, of object displacement, the matrix $\mathbf{R}$ corresponds to rotation and the column $\mathbf{p}$ corresponds to translation of the object. We shall examine both cases through simple examples. Let us first clear up the meaning of the homogenous transformation matrix describing the pose of an arbitrary frame with respect to the reference frame. Let us consider the following product of homogenous matrices which gives a new homogenous transformation matrix $\mathbf{H}$

$$\begin{aligned} \mathbf{H} &= Trans(4,-3,7)Rot(y,90^\circ)Rot(z,90^\circ) \\ &= \begin{bmatrix} 1 & 0 & 0 & 4 \\ 0 & 1 & 0 & -3 \\ 0 & 0 & 1 & 7 \\ 0 & 0 & 0 & 1 \end{bmatrix} \begin{bmatrix} 0 & 0 & 1 & 0 \\ 0 & 1 & 0 & 0 \\ -1 & 0 & 0 & 0 \\ 0 & 0 & 0 & 1 \end{bmatrix} \begin{bmatrix} 0 & -1 & 0 & 0 \\ 1 & 0 & 0 & 0 \\ 0 & 0 & 1 & 0 \\ 0 & 0 & 0 & 1 \end{bmatrix} \\ &= \begin{bmatrix} 0 & 0 & 1 & 4 \\ 1 & 0 & 0 & -3 \\ 0 & 1 & 0 & 7 \\ 0 & 0 & 0 & 1 \end{bmatrix}. \end{aligned} \tag{2.10}$$

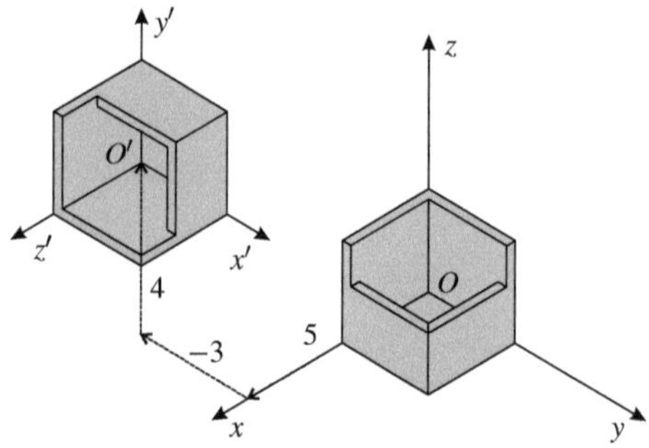

Fig. 2.8 The pose of an arbitrary frame $[x' \; y' \; z']$ with respect to the reference frame $[x \; y \; z]$

When defining the homogenous matrix representing rotation, we learned that the first three columns describe the rotation of the frame x', y', z' with respect to the reference frame x, y, z

$$\begin{array}{c} \begin{matrix} x' & y' & z' \end{matrix} \\ \begin{bmatrix} \begin{bmatrix} 0 \\ 1 \\ 0 \end{bmatrix} & \begin{bmatrix} 0 \\ 0 \\ 1 \end{bmatrix} & \begin{bmatrix} 1 \\ 0 \\ 0 \end{bmatrix} & \begin{matrix} 4 \\ -3 \\ 7 \end{matrix} \\ 0 & 0 & 0 & 1 \end{bmatrix} \begin{matrix} x \\ y \\ z \\ \; \end{matrix} \end{array} . \tag{2.11}$$

The fourth column represents the position of the origin of the frame x', y', z' with respect to the reference frame x, y, z. With this knowledge we can graphically represent the frame x', y', z' described by the homogenous transformation matrix (2.10), relative to the reference frame x, y, z (Fig. 2.8). The x' axis points in the direction of y axis of the reference frame, the y' axis is in the direction of the z axis, and the z' axis is in the x direction.

To convince ourselves of the correctness of the frame drawn in Fig. 2.8, we shall check the displacements included in Eq. (2.10). The reference frame is first translated into the point $[4, -3, 7]^T$. It is then rotated for 90° around the new y axis and finally it is rotated for 90° around the newest z axis (Fig. 2.9). The three displacements of the reference frame result in the same final pose as shown in Fig. 2.8.

In the continuation of this chapter we wish to elucidate the second meaning of the homogenous transformation matrix, i.e. a displacement of an object or coordinate frame into a new pose (Fig. 2.10). First, we wish to rotate the coordinate frame x, y, z by 90° in the counter-clockwise direction around the z axis. This can be achieved by the following postmultiplication of the matrix $\mathbf{H}$ describing the initial pose of the coordinate frame x, y, z

$$\mathbf{H}_1 = \mathbf{H} \cdot Rot(z, 90°). \tag{2.12}$$

The displacement results in a new pose of the object and new frame x', y', z' shown in Figure 2.10. We shall displace this new frame by -1 along the x' axis, 3 units

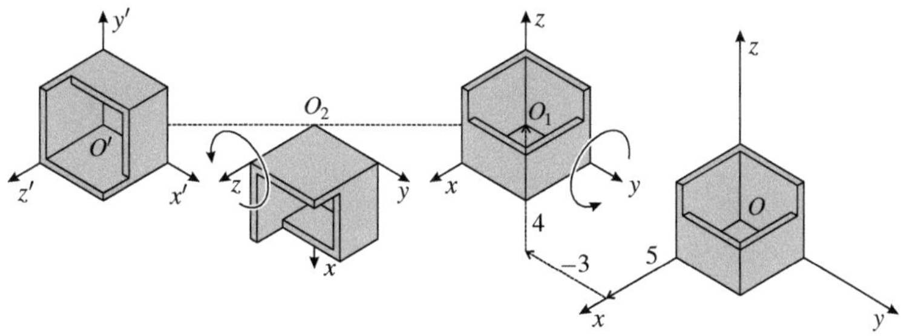

Fig. 2.9 Displacement of the reference frame into a new pose (from *right* to *left*). The origins O_1, O_2 and O' are in the same point

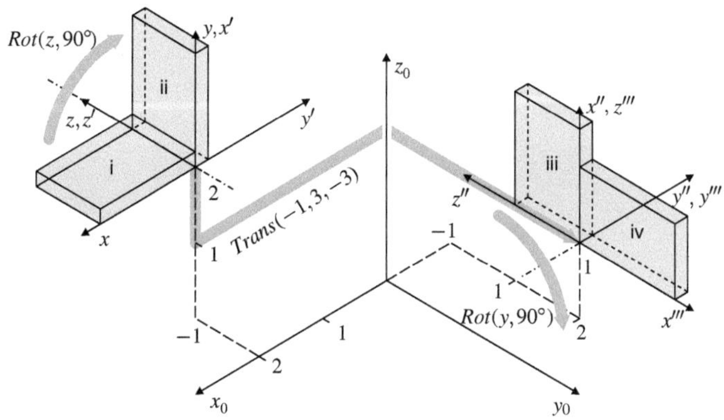

Fig. 2.10 Displacement of the object into a new pose

along y' axis and -3 along z' axis

$$\mathbf{H}_2 = \mathbf{H}_1 \cdot Trans(-1, 3, -3). \tag{2.13}$$

After translation a new pose of the object is obtained together with a new frame x'', y'', z''. This frame will be finally rotated for 90° around the y'' axis in the positive direction

$$\mathbf{H}_3 = \mathbf{H}_2 \cdot Rot(y'', 90°). \tag{2.14}$$

The Eqs. (2.12), (2.13) and (2.14) can be successively inserted one into another

$$\mathbf{H}_3 = \mathbf{H} \cdot Rot(z, 90°) \cdot Trans(-1, 3, -3) \cdot Rot(y'', 90°) = \mathbf{H} \cdot \mathbf{D}. \tag{2.15}$$

In Eq. (2.15) the matrix $\mathbf{H}$ represents the initial pose of the frame, $\mathbf{H}_3$ is the final pose, while $\mathbf{D}$ represents the displacement

$$\mathbf{D} = Rot(z, 90°) \cdot Trans(-1, 3, -3) \cdot Rot(y'', 90°)$$
$$= \begin{bmatrix} 0 & -1 & 0 & 0 \\ 1 & 0 & 0 & 0 \\ 0 & 0 & 1 & 0 \\ 0 & 0 & 0 & 1 \end{bmatrix} \begin{bmatrix} 1 & 0 & 0 & -1 \\ 0 & 1 & 0 & 3 \\ 0 & 0 & 1 & -3 \\ 0 & 0 & 0 & 1 \end{bmatrix} \begin{bmatrix} 0 & 0 & 1 & 0 \\ 0 & 1 & 0 & 0 \\ -1 & 0 & 0 & 0 \\ 0 & 0 & 0 & 1 \end{bmatrix} \tag{2.16}$$
$$= \begin{bmatrix} 0 & -1 & 0 & -3 \\ 0 & 0 & 1 & -1 \\ -1 & 0 & 0 & -3 \\ 0 & 0 & 0 & 1 \end{bmatrix}.$$

Finally we shall perform the postmultiplication describing the new relative pose of the object

$$\mathbf{H}_3 = \mathbf{H} \cdot \mathbf{D} = \begin{bmatrix} 1 & 0 & 0 & 2 \\ 0 & 0 & -1 & -1 \\ 0 & 1 & 0 & 2 \\ 0 & 0 & 0 & 1 \end{bmatrix} \begin{bmatrix} 0 & -1 & 0 & -3 \\ 0 & 0 & 1 & -1 \\ -1 & 0 & 0 & -3 \\ 0 & 0 & 0 & 1 \end{bmatrix}$$
$$= \begin{matrix} & x''' & y''' & z''' & \\ \end{matrix} \tag{2.17}$$
$$= \begin{bmatrix} 0 & -1 & 0 & -1 \\ 1 & 0 & 0 & 2 \\ 0 & 0 & 1 & 1 \\ 0 & 0 & 0 & 1 \end{bmatrix} \begin{matrix} x_0 \\ y_0 \\ z_0 \\ \end{matrix}.$$

As in the previous example, we shall graphically verify the correctness of the matrix (2.17). The three displacements of the frame x, y, z: rotation for 90° in counter-clockwise direction around the z axis, translation for -1 along the x' axis, 3 units along y' axis and -3 along z' axis, and rotation for 90° around y'' axis in the positive direction are shown in Fig. 2.10. The result is the final pose of the object x''', y''', z'''. The x''' axis points in the positive direction of the y_0 axis, y''' points in the negative direction of x_0 axis and z''' points in the positive direction of z_0 axis of the reference frame. The directions of the axes of the final frame correspond to the first three columns of the matrix $\mathbf{H}_3$. There is also agreement between the position of the origin of the final frame in Fig. 2.10 and the fourth column of the matrix $\mathbf{H}_3$.

2.4.1 Euler Angles

The rotation matrix $\mathbf{R}$ represents a redundant description of orientation. Orientation of an object in space is determined by three parameters while the rotation matrix consists of nine values. A minimal representation of orientation with three parameters can be achieved with the use of three rotation angles $\boldsymbol{\phi} = \begin{bmatrix} \varphi & \vartheta & \psi \end{bmatrix}^T$. We first assume a rotation matrix that describes rotation around a single coordinate frame

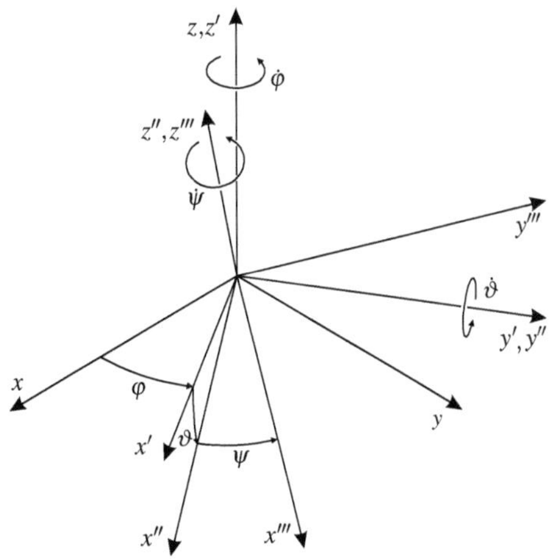

Fig. 2.11 Representation of orientation with *ZYZ* Euler angles

axis as a function of one of the three angles. A general rotation around three axes can be obtained as a combination of three consecutive rotations where two consecutive rotations must not be executed around two parallel axes. A representation of orientation of object in space can be achieved with 12 different combinations of the three elementary rotations around coordinate frame axes (for example, the combination *ZYZ* indicates first rotation around the z axis, then rotation around the y axis of already displaced coordinate frame and finally rotation around z axis of a coordinate frame that was already displaced twice beforehand; relations are shown in Fig. 2.11). Each such sequence of rotations represents a triad of Euler angles.

Rotation *ZYZ* is defined as a sequence of the following elementary rotations (Fig. 2.11):

- Rotation of the initial coordinate frame by angle φ around z axis. Rotation is determined by the matrix $\mathbf{R}_z(\varphi)$.
- Rotation of the current coordinate frame by angle ϑ around y' axis. Rotation is determined by matrix $\mathbf{R}_{y'}(\vartheta)$.
- Rotation of the current coordinate frame by angle ψ around z'' axis. Rotation is determined by matrix $\mathbf{R}_{z''}(\psi)$.

The resulting overall rotation is determined by the following sequence of elementary rotations

$$\begin{aligned}\mathbf{R}(\boldsymbol{\phi}) &= \mathbf{R}_z(\varphi)\mathbf{R}_{y'}(\vartheta)\mathbf{R}_{z''}(\psi)\\ &= \begin{bmatrix} c_\varphi c_\vartheta c_\psi - s_\varphi s_\psi & -c_\varphi c_\vartheta s_\psi - s_\varphi c_\psi & c_\varphi s_\vartheta \\ s_\varphi c_\vartheta c_\psi + c_\varphi s_\psi & -s_\varphi c_\vartheta s_\psi + c_\varphi c_\psi & s_\varphi s_\vartheta \\ -s_\vartheta c_\psi & s_\vartheta s_\psi & c_\vartheta \end{bmatrix} = \begin{bmatrix} r_{11} & r_{12} & r_{13} \\ r_{21} & r_{22} & r_{23} \\ r_{31} & r_{32} & r_{33} \end{bmatrix}\end{aligned} \tag{2.18}$$

If elements of matrix $\mathbf{R}(\boldsymbol{\phi})$ are known, Euler angles can be computed. Assuming that $r_{13} \neq 0$ and $r_{23} \neq 0$, the angle φ can be computed as

$$\varphi = \text{atan2}(r_{23}, r_{13}), \tag{2.19}$$

where $atan2$ is the four-quadrant arctangent function. By squaring and summing elements r_{13} and r_{23} and by considering r_{33}, we obtain

$$\vartheta = \text{atan2}\left(\sqrt{r_{13}^2 + r_{23}^2}, r_{33}\right). \tag{2.20}$$

The choice of positive sign for term $\sqrt{r_{13}^2 + r_{23}^2}$ constrains the value of angle ϑ to $(0, \pi)$. Angle ψ can be computed from the equation

$$\psi = \text{atan2}(r_{32}, -r_{31}). \tag{2.21}$$

Euler angles cannot be computed in poses where the arctangent function does not return a real value, Euler angles cannot be computed. These are representational singular values that depend on the selected sequence of Euler elementary rotations.

2.4.2 *Quaternions*

A quaternion is determined by a quadruple of real numbers $q = [w, x, y, z]$, for which the operations of addition and multiplication are defined by special rules. Quaternions are generalizations of complex numbers. Complex numbers enable operations with two-dimensional vectors while quaternions enable operations with four-dimensional vectors. Elements of a quaternion can be grouped into a real part w (scalar value) and imaginary part $\mathbf{v} = (x, y, z)$ (vector). A quaternion can thus be represented as $q = [w, \mathbf{v}]$. A unit quaternion has a norm of 1.

Multiplication of quaternions is determined as

$$q_1 q_2 = \begin{bmatrix} w_1 \\ \mathbf{v}_1 \end{bmatrix} \begin{bmatrix} w_2 \\ \mathbf{v}_2 \end{bmatrix} = \begin{bmatrix} w_1 w_2 - \mathbf{v}_1 \mathbf{v}_2 \\ w_1 \mathbf{v}_2 + w_2 \mathbf{v}_1 + \mathbf{v}_1 \times \mathbf{v}_2 \end{bmatrix}. \tag{2.22}$$

A conjugated quaternion q^* is defined as

$$q^* = \begin{bmatrix} w \\ \mathbf{v} \end{bmatrix}^* = \begin{bmatrix} w \\ -\mathbf{v} \end{bmatrix}, \tag{2.23}$$

and the quaternion inverse can be written as

$$q^{-1} = \frac{q^*}{|q|}, \tag{2.24}$$

where $|q|$ is the quaternion norm.

A vector (point) $\mathbf{p}$ in three-dimensional space can be written in a quaternion form as $q_\mathbf{p} = [0, \mathbf{p}]$. With the introduction of quaternion $q = [\cos\frac{\vartheta}{2}, \sin\frac{\vartheta}{2}\mathbf{s}]$ it is possible to rotate vector $\mathbf{p}$ around an arbitrary axis $\mathbf{s}$ by angle ϑ with the multiplication of quaternions

$$q q_\mathbf{p} q^{-1} = \begin{bmatrix} \cos\frac{\vartheta}{2} \\ \sin\frac{\vartheta}{2}\mathbf{s} \end{bmatrix} \begin{bmatrix} 0 \\ \mathbf{p} \end{bmatrix} \begin{bmatrix} \cos\frac{\vartheta}{2} \\ -\sin\frac{\vartheta}{2}\mathbf{s} \end{bmatrix}. \tag{2.25}$$

Representation of rotation with a quaternion requires a quadruple of real numbers. It is trivial to define a quaternion that corresponds to a rotation axis and the angle of rotation. Based on the use of quaternions, it is possible to achieve smooth rotations with quaternion interpolation. Quaternions are often used in computer graphics, as they require less parameters than rotation matrices and are not susceptible to singularities as is the case with Euler angles.

2.5 Pose of Elements in Mechanical Assembly

The knowledge of describing the pose of an object by the use of homogenous transformation matrices will first be applied to a mechanical assembly. A mechanical assembly in a virtual environment can represent a model of a device with movable parts,an avatar or an arbitrary assembly of coupled objects. For this purpose a mechanical assembly consisting of four blocks, as shown in Fig. 2.12, will be considered. A plate with dimensions $(5 \times 15 \times 1)$ is placed over a block $(5 \times 4 \times 10)$. Another plate $(8 \times 4 \times 1)$ is positioned perpendicularly to the first one, holding another small block $(1 \times 1 \times 5)$. Elements of the assembly are connected in series. This means that a displacement of one element will result in displacement of all elements that are above the displaced element in the chain and are directly or indirectly attached to the displaced element.

A frame is attached to each of the four blocks as shown in Fig. 2.12. Our task is to calculate the pose of the O_3 frame with respect to the reference frame O_0. In the last chapters we learned that the pose of a displaced frame can be expressed with respect to the reference frame by the use of the homogenous transformation matrix $\mathbf{H}$. The pose of the frame O_1 with respect to the frame O_0 is denoted as ${}^0\mathbf{H}_1$. In the same way ${}^1\mathbf{H}_2$ represents the pose of O_2 frame with respect to O_1 and ${}^2\mathbf{H}_3$ represents the pose of O_3 with regard to O_2 frame. We learned also that the successive displacements are expressed by postmultiplications (successive multiplications from left to right) of homogenous transformation matrices. Also the assembly process can be described by postmultiplication of the corresponding matrices. The pose of the fourth block can be written with respect to the first one by the following matrix

$${}^0\mathbf{H}_3 = {}^0\mathbf{H}_1 {}^1\mathbf{H}_2 {}^2\mathbf{H}_3. \tag{2.26}$$

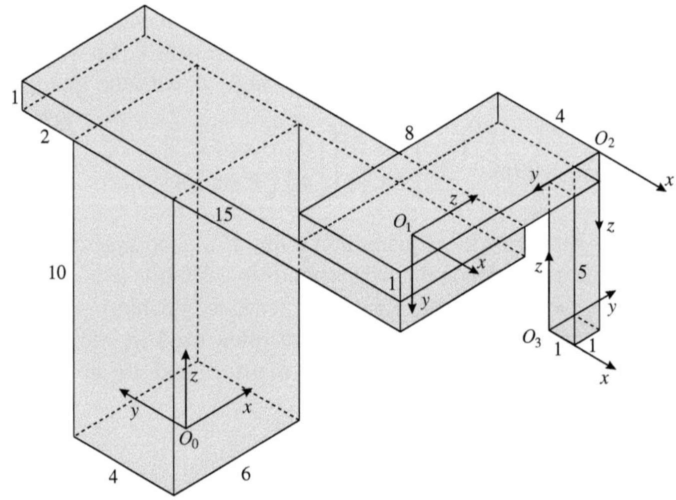

Fig. 2.12 Mechanical assembly

The blocks are positioned perpendicularly one to another. In this way it is not necessary to calculate the sines and cosines of the rotation angles. The matrices can be determined directly from Fig. 2.12. The x axis of frame O_1 points in negative direction of the y axis in the frame O_0. The y axis of frame O_1 points in negative direction of the z axis in the frame O_0. The z axis of the frame O_1 has the same direction as x axis of the frame O_0. The described geometrical properties of the assembly structure are written into the first three columns of the homogenous matrix. The position of the origin of the frame O_1 with respect to the frame O_0 is written into the fourth column

$$
{}^{0}\mathbf{H}_1 = \begin{array}{c} \overbrace{\begin{array}{ccc} x & y & z \end{array}}^{O_1} \\ \left[\begin{array}{cccc} 0 & 0 & 1 & 0 \\ -1 & 0 & 0 & 6 \\ 0 & -1 & 0 & 11 \\ 0 & 0 & 0 & 1 \end{array}\right] \end{array} \begin{array}{c} \\ \left.\begin{array}{c} x \\ y \\ z \\ \end{array}\right\} O_0 \end{array}. \tag{2.27}
$$

The other two matrices are determined in the same way

$$
{}^{1}\mathbf{H}_2 = \begin{bmatrix} 1 & 0 & 0 & 11 \\ 0 & 0 & 1 & -1 \\ 0 & -1 & 0 & 8 \\ 0 & 0 & 0 & 1 \end{bmatrix} \tag{2.28}
$$

$$
{}^{2}\mathbf{H}_{3} = \begin{bmatrix} 1 & 0 & 0 & 3 \\ 0 & -1 & 0 & 1 \\ 0 & 0 & -1 & 6 \\ 0 & 0 & 0 & 1 \end{bmatrix}. \tag{2.29}
$$

The position and orientation of the fourth block with respect to the first one are given by the ${}^{0}\mathbf{H}_{3}$ matrix, which is obtained by successive multiplication of the matrices (2.27), (2.28) and (2.29)

$$
{}^{0}\mathbf{H}_{3} = \begin{bmatrix} 0 & 1 & 0 & 7 \\ -1 & 0 & 0 & -8 \\ 0 & 0 & 1 & 6 \\ 0 & 0 & 0 & 1 \end{bmatrix}. \tag{2.30}
$$

The fourth column of the matrix ${}^{0}\mathbf{H}_{3}$ $[7, -8, 6, 1]^T$ represents the position of the origin of the frame O_3 with respect to the reference frame O_0. The correctness of the fourth column can be checked from Fig. 2.12. The rotational part of the matrix ${}^{0}\mathbf{H}_{3}$ represents the orientation of the frame O_3 with respect to the reference frame O_0.

Now let us imagine that the first horizontal plate rotates with respect to the first vertical block around axis 1 (z axis of coordinate frame O_0) for angle ϑ_1. The second plate also rotates around the vertical axis 2 (y axis of coordinate frame O_1) for angle ϑ_2. The last block is elongated by distance d_3 along the third axis (z axis of coordinate frame O_2). The new pose of the mechanism is shown in Fig. 2.13.

Since we introduced motion between the elements of the mechanism, the transformation between two consecutive blocks now consists of the matrix $\mathbf{D}_i$ that defines translational or rotational movement and matrix ${}^{i-1}\mathbf{H}_i$ that defines pose of a block.

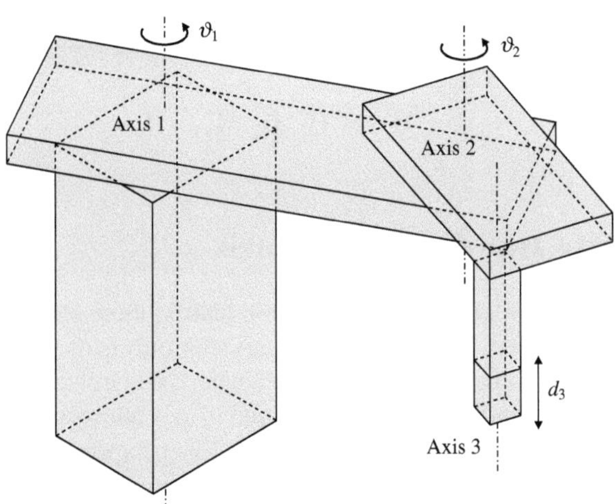

Fig. 2.13 Displacements of the mechanical assembly

Based on the definition of coordinate frames in Fig. 2.12, matrix $\mathbf{D}_1$ represents rotation around z_0 axis. Thus, the following matrix product determines the pose of frame O_1 relative to the frame O_0 (pose of the second block relative to the first block)

$$
{}^0\mathbf{H}_1' = \mathbf{D}_1{}^0\mathbf{H}_1 = \begin{bmatrix} \cos\vartheta_1 & -\sin\vartheta_1 & 0 & 0 \\ \sin\vartheta_1 & \cos\vartheta_1 & 0 & 0 \\ 0 & 0 & 1 & 0 \\ 0 & 0 & 0 & 1 \end{bmatrix} \begin{bmatrix} 0 & 0 & 1 & 0 \\ -1 & 0 & 0 & -9 \\ 0 & -1 & 0 & 11 \\ 0 & 0 & 0 & 1 \end{bmatrix}.
$$

The second rotation is accomplished around the y_1 axis. The following matrix product defines the pose of coordinate frame O_2 relative to frame O_1 (pose of the third block relative to the second block)

$$
{}^1\mathbf{H}_2' = \mathbf{D}_2{}^1\mathbf{H}_2 = \begin{bmatrix} \cos\vartheta_2 & 0 & \sin\vartheta_2 & 0 \\ 0 & 1 & 0 & 0 \\ -\sin\vartheta_2 & 0 & \cos\vartheta_2 & 0 \\ 0 & 0 & 0 & 1 \end{bmatrix} \begin{bmatrix} 1 & 0 & 0 & 2 \\ 0 & 0 & 1 & -1 \\ 0 & -1 & 0 & 5 \\ 0 & 0 & 0 & 1 \end{bmatrix}.
$$

In the last joint, we are dealing with translation along z_2 axis (pose of the fourth block relative to the third block)

$$
{}^2\mathbf{H}_3' = \mathbf{D}_3{}^2\mathbf{H}_3 = \begin{bmatrix} 1 & 0 & 0 & 0 \\ 0 & 1 & 0 & 0 \\ 0 & 0 & 1 & d_3 \\ 0 & 0 & 0 & 1 \end{bmatrix} \begin{bmatrix} 1 & 0 & 0 & -1 \\ 0 & -1 & 0 & 1 \\ 0 & 0 & -1 & 6 \\ 0 & 0 & 0 & 1 \end{bmatrix}.
$$

Matrices ${}^0\mathbf{H}_1'$, ${}^1\mathbf{H}_2'$ and ${}^2\mathbf{H}_3'$ determine relative poses of elements of the mechanical assembly after the completed displacements. The pose of the last block (frame O_3) relative to the first block (frame O_0) can be computed as a postmultiplication of matrices ${}^{i-1}\mathbf{H}_i'$

$$
{}^0\mathbf{H}_3' = {}^0\mathbf{H}_1'{}^1\mathbf{H}_2'{}^2\mathbf{H}_3'. \tag{2.31}
$$

2.6 Perspective Transformation Matrix

When defining the homogenous transformation matrix, three zeros and a one were written into the fourth line. Their purpose appears to be only to make the homogenous matrix quadratic. In this chapter we shall learn that the last line of the matrix represents the perspective. The homogenous transformation matrix allows us not only to move an object, but also to deform it or scale it in size. The perspective transformation is relevant for computer graphics and design of virtual environments. The perspective transformation can be explained by formation of the image of an object through the lens with focal length f (Fig. 2.14). The lens equation is

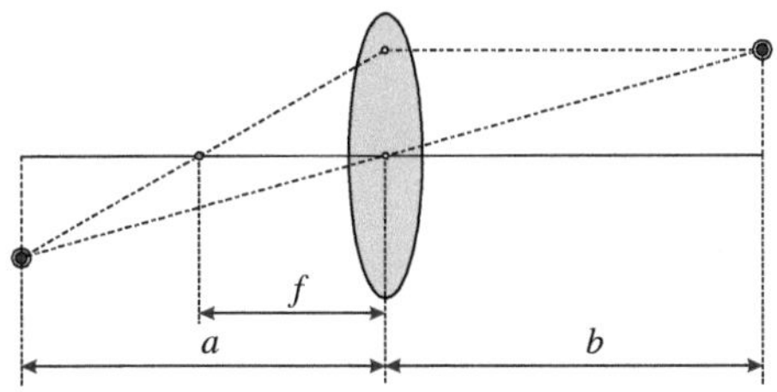

Fig. 2.14 Image formation by the lens

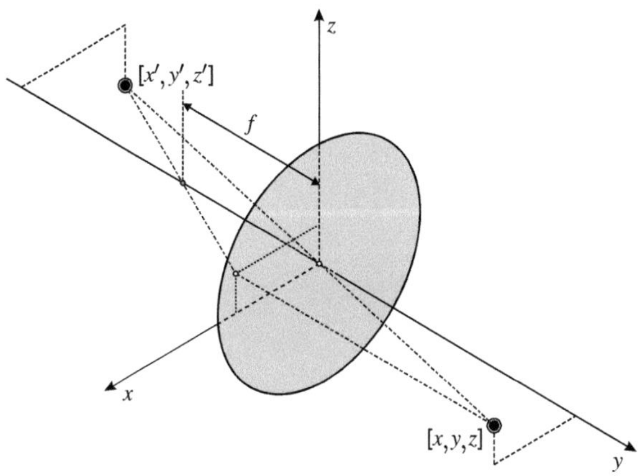

Fig. 2.15 Image of a point through the lens

$$\frac{1}{a} + \frac{1}{b} = \frac{1}{f}. \tag{2.32}$$

Let us place the lens into the x, z plane of the Cartesian coordinate frame (Fig. 2.15). The point with coordinates $[x, y, z]^{\mathrm{T}}$ is imaged into the point $[x', y', z']^{T}$. The lens equation in this particular situation is as follows

$$\frac{1}{y} - \frac{1}{y'} = \frac{1}{f}. \tag{2.33}$$

The rays passing through the center of the lens remain undeviated

$$\frac{z}{y} = \frac{z'}{y'}. \tag{2.34}$$

Another equation for undeviated rays is obtained by exchanging z and z' with x and x' in Eq. (2.34). When rearranging the equations for deviated and undeviated rays, we can obtain the relations between the coordinates of the original point x, y, and z and its image x', y', z'

$$x' = \frac{x}{1 - \frac{y}{f}}, \tag{2.35}$$

$$y' = \frac{y}{1 - \frac{y}{f}}, \tag{2.36}$$

$$z' = \frac{z}{1 - \frac{y}{f}}. \tag{2.37}$$

The same result is obtained by the use of the homogenous matrix **P**, which describes the perspective transformation

$$\mathbf{P} = \begin{bmatrix} 1 & 0 & 0 & 0 \\ 0 & 1 & 0 & 0 \\ 0 & 0 & 1 & 0 \\ 0 & -\frac{1}{f} & 0 & 1 \end{bmatrix}. \tag{2.38}$$

The coordinates of the imaged point x', y', z' are obtained by multiplying the coordinates of the original point x, y, z by the matrix **P**

$$w' \begin{bmatrix} x' \\ y' \\ z' \\ 1 \end{bmatrix} = \begin{bmatrix} 1 & 0 & 0 & 0 \\ 0 & 1 & 0 & 0 \\ 0 & 0 & 1 & 0 \\ 0 & -\frac{1}{f} & 0 & 1 \end{bmatrix} \begin{bmatrix} x \\ y \\ z \\ 1 \end{bmatrix} = \begin{bmatrix} x \\ y \\ z \\ 1 - \frac{y}{f} \end{bmatrix}, \tag{2.39}$$

where w' is a scaling factor. The same relation between the imaged and original coordinates was obtained as in Eqs. (2.35–2.37). When the element $-\frac{1}{f}$ is at the bottom of the first column, we are dealing with a perspective transformation along the x axis. When it is at the bottom of the third column, we have projection along the z axis.

References

1. Bajd T, Mihelj M, Lenarčič J, Stanovnik A, Munih M (2010) Robotics. Springer
2. Bajd T, Mihelj M, Munih M (2013) Introduction to Robotics. Springer

Chapter 3
Dynamic Model of a Virtual Environment

Abstract A virtual world represents an environment consisting of a variety of objects, avatars, actuators and other elements. In general, we are dealing with dynamic environments where objects can move under forces and torques originating from various sources. In this chapter the basic mathematical tools in the study of rigid body dynamics will be introduced. These include equations of motion, rigid body inertial properties, linear and angular momentum and collision detection between rigid bodies. Finally, a closed loop algorithm for computation of body motion is provided.

A virtual world represents an environment consisting of a variety of objects, avatars, actuators and other elements. In general, we are dealing with dynamic environments where objects can move when touched. Forces and torques from various sources act on the virtual objects. Simulating the dynamics of virtual environments is an important part of an application, regardless of whether it is dynamics of rigid bodies, deformation dynamics or dynamics of fluids. Inclusion of the relevant dynamic responses allows realistic behavior of the virtual environment to be achieved.

A model of a body in a virtual environment must include a description of its dynamic behavior. This description also defines the body's physical interaction with other bodies in the environment. Body dynamics can be described based on various assumptions, which then determine the level of realism and the computational complexity of simulation.

A static environment, for example, consists of stationary objects around which the user moves. Real-world physical laws are not implemented in the virtual environment. In this case, the computational complexity is the lowest.

On the other hand, Newtonian physics represents an excellent approximation of real-world physics and includes conservation of momentum as well as action and reaction forces. Objects behave realistically, but computational complexity increases significantly. This can be simplified by a set of rules that are less accurate than Newtonian physics, but often describe developments in a way that seems natural to most humans.

M. Mihelj et al., *Virtual Reality Technology and Applications*,
Intelligent Systems, Control and Automation: Science and Engineering 68,
DOI: 10.1007/978-94-007-6910-6_3,

Newtonian physics can be upgraded with physical laws that describe events in an environment that is beyond our perceptions. These laws apply either to micro (molecules, atoms) or macro environments (galaxies, universe) and are defined by quantum and relativistic physics.

The basic concepts applicable to Newtonian physics are summarized below. A static world and simplified physics are simplifications of Newton dynamics while the concepts of non-classical physics are beyond the scope of this chapter and will not be discussed here.

3.1 Equations of Motion

Figure 3.1 shows a mass particle, which is considered a dimensionless object, and a body constructed from at least three interconnected mass particles. For the purpose of further consideration we will assume that the body is composed of exactly three mass particles. The concept can be expanded for more complex bodies. The coordinate system of the body is located in the body's center of mass. Vectors $\mathbf{r}_i$ determine the position of each mass particle relative to the body's coordinate frame.

Although both a mass particle as well as a body generally move in a three-dimensional space, we will assume movement constrained to a plane for purposes of explanation.

First we will consider the motion of a mass particle. Since we are not interested in particle orientation, the particle motion can be described using the position vector $\mathbf{p}(t)$ and its time derivatives. Velocity of a mass particle is defined as a time derivative of the position vector $\mathbf{p}(t)$ as

$$\mathbf{v}(t) = \dot{\mathbf{p}}(t) = \frac{d\mathbf{p}(t)}{dt}. \tag{3.1}$$

Knowing the motion velocity of a mass particle, its position can be computed as a time integral

$$\mathbf{p}(t) = \mathbf{p}_0 + \int_0^t \mathbf{v}(\xi)d\xi, \tag{3.2}$$

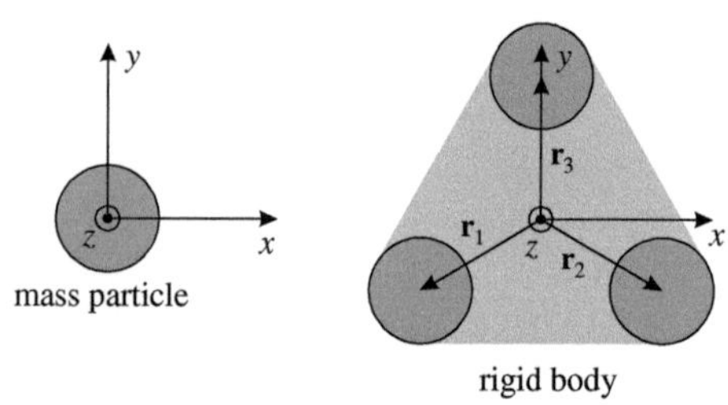

Fig. 3.1 Concept of a mass particle and a rigid body

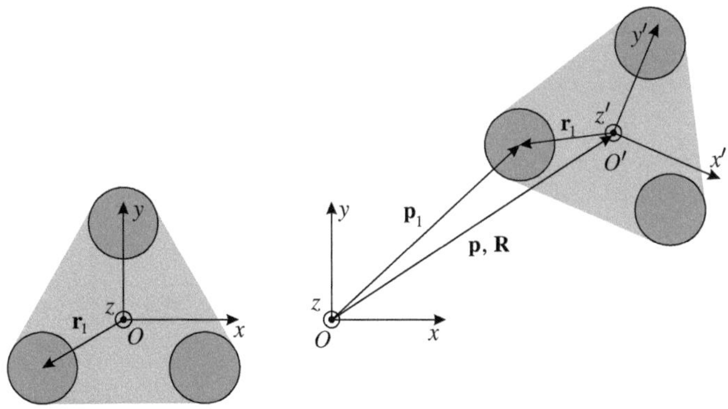

Fig. 3.2 Displacement of a rigid body from an initial pose (*left*) to a final pose (*right*)

Fig. 3.3 Motion of a rigid body

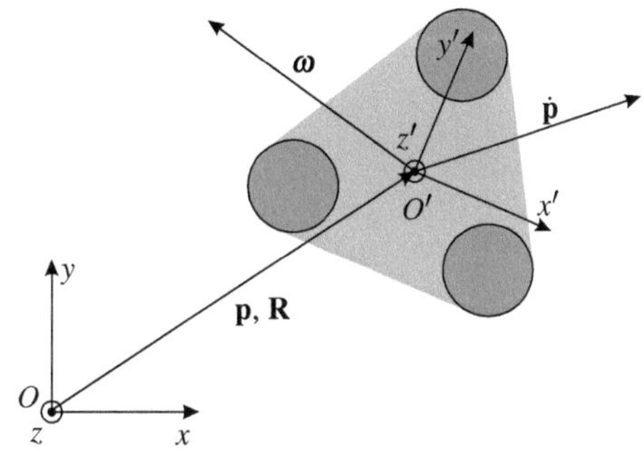

where $\mathbf{p}_0$ indicates the initial position.

The two equations above are also valid for a rigid body. However, in this case we need to consider also the body orientation. Their relations are shown in Fig. 3.2. Vector $\mathbf{p}$ and matrix $\mathbf{R}$ determine the position and orientation of the rigid body relative to the coordinate frame O_0. The body orientation can also be written in terms of quaternion q.

The position of a point $\mathbf{p}_1(t)$ located on the rigid body can be computed relative to the coordinate frame O_0 as

$$\mathbf{p}_1(t) = \mathbf{p}(t) + \mathbf{R}(t)\mathbf{r}_1, \tag{3.3}$$

or

$$\begin{bmatrix} \mathbf{p}_1(t) \\ 1 \end{bmatrix} = \begin{bmatrix} \mathbf{R}(t) & \mathbf{p}(t) \\ \mathbf{0} & 1 \end{bmatrix} \begin{bmatrix} \mathbf{r}_1 \\ 1 \end{bmatrix} = \mathbf{T}(t) \begin{bmatrix} \mathbf{r}_1 \\ 1 \end{bmatrix}, \tag{3.4}$$

where $\mathbf{T}(t)$ represents a homogenous transformation matrix.

During the motion of a rigid body, we need to consider translational as well as rotational velocities and accelerations. Figure 3.3 shows the two velocity vectors.

The translational velocity is determined as a time derivative defined in Eq. (3.1) of the position vector that determines the body's center of mass. The change of orientation leads to

$$\dot{\mathbf{R}}(t) = \boldsymbol{\omega}^*(t)\mathbf{R}(t), \tag{3.5}$$

when body orientation is determined in terms of a rotation matrix or

$$\dot{q}(t) = \frac{1}{2}\tilde{\omega}(t) \otimes q(t), \tag{3.6}$$

when body orientation is determined in terms of a quaternion. Matrix $\boldsymbol{\omega}^*(t)$ is a skew-symmetric matrix of vector $\boldsymbol{\omega}(t) = \left[\omega_x(t)\ \omega_y(t)\ \omega_z(t)\right]^T$, which determines the body's angular velocity, such that

$$\boldsymbol{\omega}^*(t) = \begin{bmatrix} 0 & -\omega_z(t) & \omega_y(t) \\ \omega_z(t) & 0 & \omega_x(t) \\ -\omega_y(t) & \omega_x(t) & 0 \end{bmatrix}. \tag{3.7}$$

Quaternion $\tilde{\omega}(t)$ is an augmented angular velocity vector $\boldsymbol{\omega}(t) = [\omega_x(t)\ \omega_y(t)\ \omega_z(t)]^T$ and operator $\otimes$ denotes quaternion multiplication.

A body's orientation in space can be computed with time integration of Eqs. (3.5) or (3.6).

3.2 Mass, Center of Mass and Moment of Inertia

Body dynamic properties depend on its mass and inertia. Body mass is defined as a sum of masses of the individual particles constituting the body (Fig. 3.4)

$$M = \sum_{i=1}^{N} m_i, \tag{3.8}$$

where N is the number of mass particles (in our case $N = 3$). Since bodies are usually composed of homogenously distributed matter and not discrete particles, the sum in the above equation should be replaced with the integral across the body volume in real conditions.

The definition of a body's center of mass enables us to separate translational and rotational dynamics. The body center of mass in the local coordinate frame can be computed as

$$\mathbf{r}_c = \frac{\sum_{i=1}^{N} m_i \mathbf{r}_i}{M}. \tag{3.9}$$

Since the local coordinate frame is positioned in the body's center of mass, the coordinates of the body center of mass expressed in the local coordinate frame equal

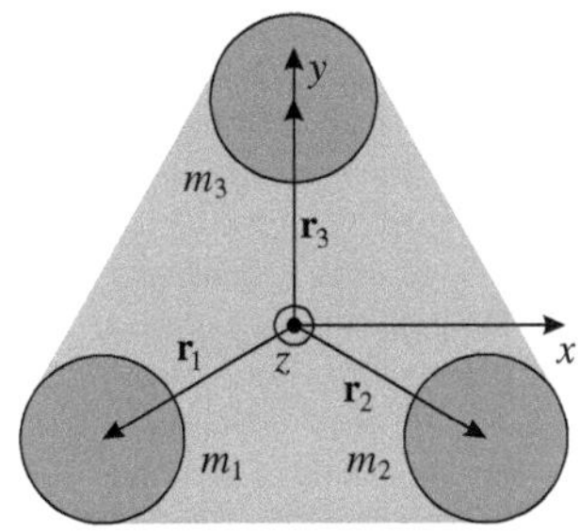

Fig. 3.4 Mass, center of mass and moment of inertia of a rigid body

zero. The body center of mass expressed in the global coordinate frame can be computed based on the Eq. (3.3).

Finally we define the body inertia tensor $\mathbf{I}_0$ with respect to the local coordinate frame. A body inertia tensor provides information about the distribution of body mass relative to the body center of mass

$$\mathbf{I}_0 = \sum_i^N \begin{bmatrix} m_i(r_{i_y}^2 + r_{i_z}^2) & -m_i r_{i_x} r_{i_y} & -m_i r_{i_x} r_{i_z} \\ -m_i r_{i_x} r_{i_y} & m_i(r_{i_x}^2 + r_{i_z}^2) & -m_i r_{i_y} r_{i_z} \\ -m_i r_{i_x} r_{i_z} & -m_i r_{i_y} r_{i_z} & m_i(r_{i_x}^2 + r_{i_y}^2) \end{bmatrix}, \tag{3.10}$$

where $\mathbf{r}_i = \left[r_{i_x} \; r_{i_y} \; r_{i_z} \right]^T$. If the body's shape does not change, the inertia tensor $\mathbf{I}_0$ is constant. The inertia tensor with respect to the global coordinate frame can be computed as

$$\mathbf{I}(t) = \mathbf{R}(t)\mathbf{I}_0\mathbf{R}^T(t). \tag{3.11}$$

In general, matrix $\mathbf{I}(t)$ is time-dependent since body orientation relative to the global coordinate frame changes during the body's motion.

3.3 Linear and Angular Momentum

Linear momentum of a dimensionless particle with mass m is defined as

$$\mathbf{G}(t) = m\mathbf{v}(t). \tag{3.12}$$

If $\mathbf{v}(t)$ specifies the velocity of a rigid body's center of mass, a similar equation can also be written for a rigid body

$$\mathbf{G}(t) = M\mathbf{v}(t). \tag{3.13}$$

In this regard, a rigid body behaves similarly to a mass particle with mass M.

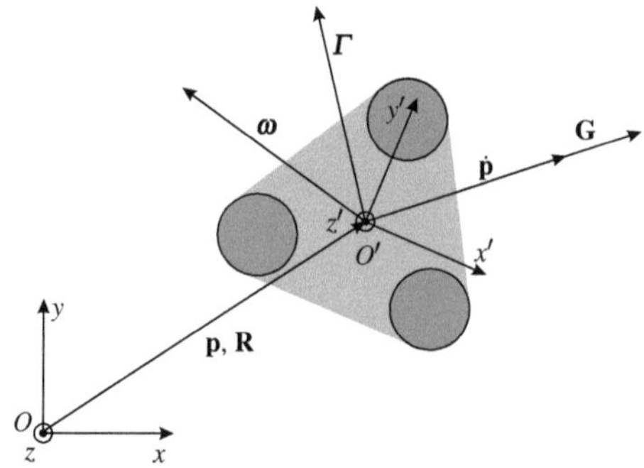

Fig. 3.5 Linear and angular momentum of a rigid body. The linear momentum vector is oriented in the direction of translational velocity. On the other hand, this is generally not the case for body angular momentum (the two vectors are aligned only if the body rotates about one of its principal axes)

If there is no external force acting on the body, the linear momentum is conserved. It is evident from Eq. (3.13) that translational velocity of the body's center of mass is also constant.

A somewhat less intuitive concept than linear momentum is the body's angular momentum (angular momentum of a dimensionless body equals zero, since its moment of inertia equals zero). Angular momentum is defined by the product

$$\boldsymbol{\Gamma}(t) = \mathbf{I}(t)\boldsymbol{\omega}(t). \tag{3.14}$$

A body's angular momentum is conserved if there is no external torque acting on the body. Conservation of angular momentum is an important reason for introducing this concept into the description of body dynamics. Namely, body rotational velocity $\omega(t)$ may change even though the angular momentum is constant. Therefore, introduction of body angular momentum simplifies computation of equations of motion compared to the use of rotational velocities.

Vectors of linear and angular momentum are shown in Fig. 3.5.

3.4 Forces and Torques Acting on a Rigid Body

Motion of a body in space depends on the forces and torques acting on it (Fig. 3.6). A force $\mathbf{F}$ acts on the body with lever $\mathbf{r}$ relative to the body's center of mass. Force $\mathbf{F}$ results in a change of the body's linear momentum. At the same time, force $\mathbf{F}$ produces torque $\mathbf{r} \times \mathbf{F}$, that, together with other possible external torques, contributes to the overall torque τ acting on the body and causes changes to the body's angular momentum.

The change in linear momentum equals the impulse of the sum of all forces $\mathbf{F}$ acting on the body, thus

$$d\mathbf{G}(t) = \mathbf{F}(t)dt. \tag{3.15}$$

The above equation can be rewritten as

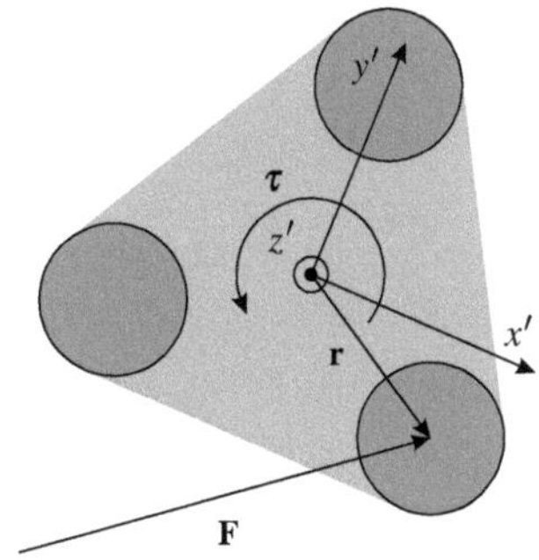

Fig. 3.6 Forces and torques acting on a rigid body

$$\frac{d\mathbf{G}(t)}{dt} = \dot{\mathbf{G}}(t) = M\dot{\mathbf{v}}(t) = \mathbf{F}(t), \tag{3.16}$$

meaning that the time derivative of linear momentum equals the product of body mass and acceleration or the sum of all forces acting on the body.

The change in angular momentum equals the impulse of the sum of all torques $\boldsymbol{\tau}$ acting on the body, thus

$$d\boldsymbol{\Gamma}(t) = \boldsymbol{\tau}(t)dt. \tag{3.17}$$

The time derivative of angular momentum thus equals the sum of all torques acting on the body

$$\frac{d\boldsymbol{\Gamma}(t)}{dt} = \dot{\boldsymbol{\Gamma}}(t) = \boldsymbol{\tau}(t). \tag{3.18}$$

If forces and torques acting on the body are known, time derivatives of linear and angular momenta are also defined. The body's linear momentum can thus be computed as a time integral of all forces acting on the body

$$\mathbf{G}(t) = \mathbf{G}_0 + \int_0^t \mathbf{F}(\xi)d\xi, \tag{3.19}$$

where $\mathbf{G}_0$ is the initial linear momentum.

Angular momentum can be computed as a time integral of all torques acting on the body

$$\boldsymbol{\Gamma}(t) = \boldsymbol{\Gamma}_0 + \int_0^t \boldsymbol{\tau}(\xi)d\xi, \tag{3.20}$$

where $\boldsymbol{\Gamma}_0$ indicates the initial angular momentum.

From the known linear $\mathbf{G}(t)$ and angular $\boldsymbol{\Gamma}(t)$ momenta as well as the body's mass and inertia properties it is possible to compute the body's translational velocity $\mathbf{v}(t)$ and rotational velocity $\boldsymbol{\omega}(t)$ based on Eqs. (3.13) and (3.14), respectively.

In the following paragraphs we will introduce some typical forces and torques acting on an body in a virtual environment. In general, these can be divided into five relevant contributions: (1) forces and torques as a result of interactions with the user (via haptic interface), (2) force field (gravity, magnetic field), (3) forces and

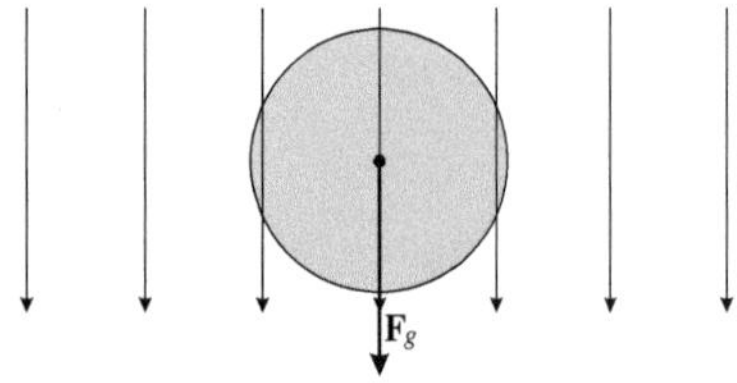

Fig. 3.7 Gravity field

torques produced by the medium through which the body moves (viscous damping), (4) forces resulting from interactions of a body with other bodies in a virtual environment (collisions) and (5) virtual actuators (sources of forces and torques).

Interaction between a user and a virtual environment is often done through virtual tools that the user manipulates through a haptic interface. Virtual actuators are sources of constant or variable forces and torques acting on a body. Virtual actuators can be models of electrical, hydraulic, pneumatic actuator systems or, for example, engines with internal combustion. This group can also include biological actuators—muscles. The magnitude of forces and torques of virtual actuators can change automatically based on events within a virtual environment or through interactions with the user.

The force field within which the bodies move can be homogeneous (local gravity field) or nonhomogeneous (magnetic dipole filed). The force acting on the body depends on the field parameters and body properties. For example, a gravity force can be computed as (Fig. 3.7)

$$\mathbf{F}_g = M\mathbf{g}, \tag{3.21}$$

where $\mathbf{g}$ is the gravity acceleration vector. In this case, the force is independent of the body motion. The homogeneous force field also does not cause any torque that would result in a change of body angular momentum.

Analysis of interaction forces between the body and the medium through which the body moves (Fig. 3.8) is relatively straightforward. In this case, friction forces are of primary interest. In the case of a simple model based on viscous damping (a body floating in a viscous fluid), the interaction force can be computed as

$$\mathbf{F}_B = -B\mathbf{v}, \tag{3.22}$$

where B is the coefficient of viscous damping and $\mathbf{v}$ indicates body velocity. Since friction forces oppose object motion, a negative sign is introduced.

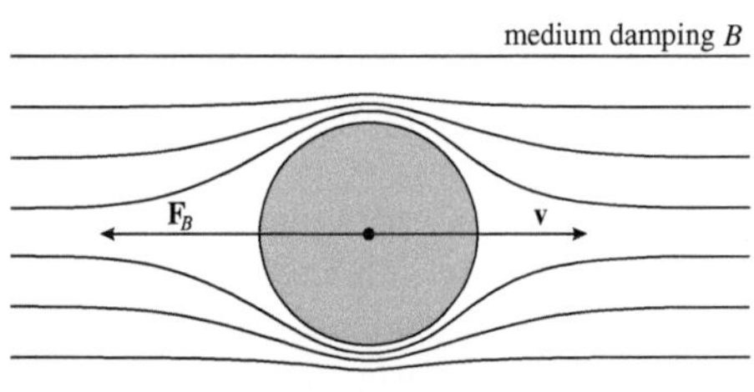

Fig. 3.8 Body in a viscous medium

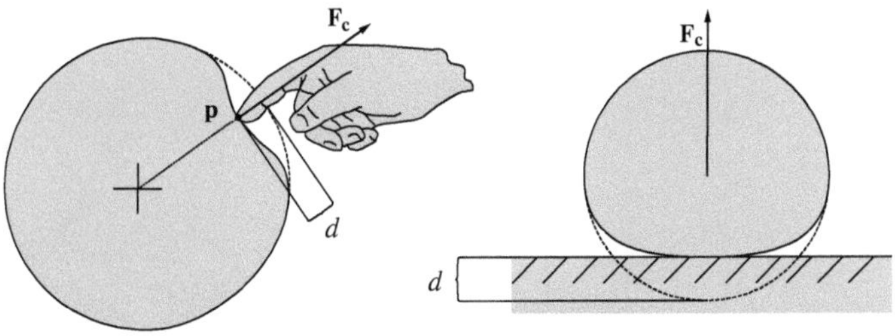

Fig. 3.9 Collision of two bodies and a collision between a body and a grounded wall; **p**—contact point, d—body deformation (penetration) and $\mathbf{F}_c$—computed reaction force

The most complex analysis is that of forces and torques resulting from interactions between bodies. A prerequisite is the implementation of an algorithm for collision detection. It is then possible to compute reaction forces based on dynamic properties (plasticity, elasticity).

Relations during a collision are shown in Fig. 3.9. As a result of a collision between two bodies, a plastic or elastic deformation occurs (the maximal deformation is indicated with d in the figure). Deformations depend on body stiffness properties. Body collisions and deformations need to be computed in real time. The computed deformations can be used to model interaction forces or for visualization purposes.

For the simplest case where a body is modeled only as a spring with stiffness value k, the collision reaction force $\mathbf{F}_c$ can be computed as

$$\mathbf{F}_c = kd\mathbf{n}, \tag{3.23}$$

where d determines the body deformation and vector $\mathbf{n}$ determines the reaction force direction. For a simple case of a frictionless contact, the vector $\mathbf{n}$ can be determined as a normal vector to the surface of the body at a point of contact. A more detailed model for computation of reaction forces occurring during collisions will be presented in Chap. 7 dealing with haptic interfaces.

3.5 Collision Detection

Equation (3.23) indicates that computation of interaction forces during collisions requires knowledge about body properties. In addition to the stiffness value, friction parameters are usually required for realistic force modeling. Furthermore, body deformation d and force direction vector $\mathbf{n}$ need to be computed from the geometry of the collision. This chapter will review some basic concepts for collision detection that are based on body geometry. In order to simplify analysis, we will mostly limit collisions to a single plane only.

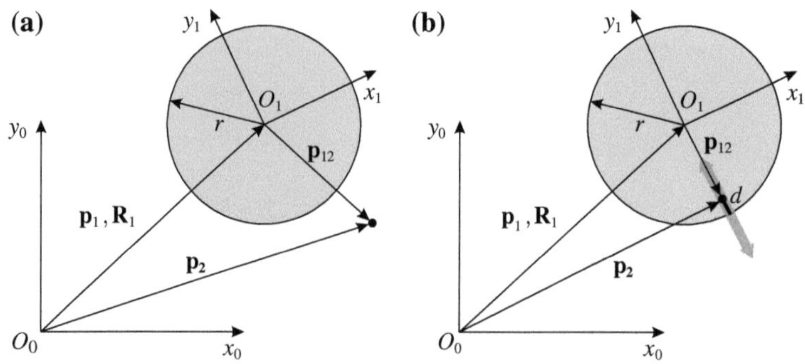

Fig. 3.10 Collision between a *sphere* and a dimensionless particle (simplified view with a collision between a *circle* and a particle); the *left* image shows relations before the collision while the *right* image shows relations after the collision. Thick *straight arrows* indicate force directions

We will first consider a collision between a sphere and a dimensionless particle (Fig. 3.10). Collision detection is relatively simple in this case. Based on relations in Fig. 3.10 it is evident that the particle collides with the sphere when the length of vector $\mathbf{p}_{12} = \mathbf{p}_2 - \mathbf{p}_1$ is smaller than the sphere radius r. Sphere deformation can be computed as

$$d = \begin{cases} 0 & \text{for} \quad \|\mathbf{p}_{12}\| > r \\ r - \|\mathbf{p}_{12}\| & \text{for} \quad \|\mathbf{p}_{12}\| < r. \end{cases} \tag{3.24}$$

In the case of a frictionless collision, the reaction force direction is determined along vector $\mathbf{p}_{12}$, which is normal to the sphere surface.

Figure 3.11 shows a collision between a block and a dimensionless particle. As in the case of collision with a sphere, the vector $\mathbf{p}_{12} = \mathbf{p}_2 - \mathbf{p}_1$ should first be computed. However, this is not sufficient since it is necessary to determine the particle position relative to individual block faces (sides of the rectangle on Fig. 3.11). Namely, vector $\mathbf{p}_{12}$ is computed relative to the global coordinate frame O_0, while block faces are generally not aligned with the axes of frame O_0. Collision detection can be simplified by transforming vector $\mathbf{p}_{12}$ into the local coordinate frame O_1, resulting in $\mathbf{p}_{12}^1$. Vector $\mathbf{p}_{12}^1$ can be computed as

$$\mathbf{p}_{12}^1 = \mathbf{R}_1^T \mathbf{p}_{12} \quad \text{or} \quad \begin{bmatrix} \mathbf{p}_{12}^1 \\ 1 \end{bmatrix} = \begin{bmatrix} \mathbf{R}_1 & \mathbf{p}_1 \\ \mathbf{0} & 1 \end{bmatrix}^{-1} \begin{bmatrix} \mathbf{p}_2 \\ 1 \end{bmatrix} = \mathbf{T}^{-1} \begin{bmatrix} \mathbf{p}_2 \\ 1 \end{bmatrix}, \tag{3.25}$$

where $\mathbf{R}_1$ is the rotation matrix that defines the orientation of the frame O_1 relative to the frame O_0. Axes of coordinate frame O_1 are aligned with block principle axes. Therefore, it becomes straightforward to verify whether a particle lies within or outside of the body's boundaries. Individual components of vector $\mathbf{p}_{12}^1$ have to be compared against block dimensions a, b and c. For relations in Fig. 3.11 it is clear that the particle lies within the rectangle's boundaries (we are considering only plane relations here) if the following condition is satisfied

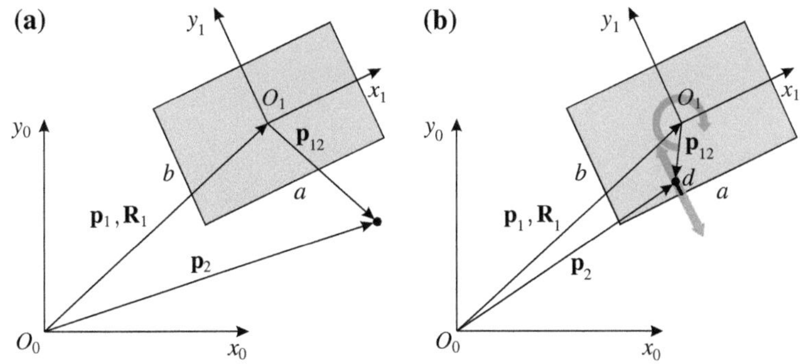

Fig. 3.11 Collision between a block and a dimensionless particle (simplified view with a collision between a *rectangle* and a particle); *left* image shows relations before the collision while the *right* image shows relations after the collision. Thick *straight arrows* indicate force directions while thick *circular arrow* indicates torque acting on the block

$$|p^1_{12_x}| < \frac{a}{2} \quad \wedge \quad |p^1_{12_y}| < \frac{b}{2}, \tag{3.26}$$

where $p^1_{12_x}$ and $p^1_{12_y}$ are the x and y components of vector $\mathbf{p}^1_{12}$.

However, in this case it is not trivial to determine deformation d and reaction force direction. Specifically, we also have to take into account the relative position and direction of motion between the block and the particle at the instant before the collision occurrence. If the collision occurred along side a (see Fig. 3.11), then the resulting deformation equals

$$d = \frac{b}{2} - |p^1_{12_y}|, \tag{3.27}$$

and the force direction for frictionless contact is along the normal vector to side a. In the opposite case

$$d = \frac{a}{2} - |p^1_{12_x}|, \tag{3.28}$$

and force direction is determined along the normal vector to the side b. Since the force vector generally does not pass through the block's center of mass, the reaction force causes an additional torque that tries to rotate the block around its center of mass. In a three-dimensional space also the third dimension should be considered in the above equations.

Transformation of a pose of one body into the local coordinate frame of the other body, as determined by Eq. (3.25), can also be used in more complex scenarios, where we have to deal with collisions between two three-dimensional bodies (it can also be used to compute a collision between a sphere and a dimensionless particle). Figure 3.12 shows a collision between two spheres. Collision analysis in this case is as simple as analysis of collision between a sphere and a particle. It is only necessary to compute vector $\mathbf{p}_{12}$. If its length is smaller than the sum of sphere radii $r_1 + r_2$,

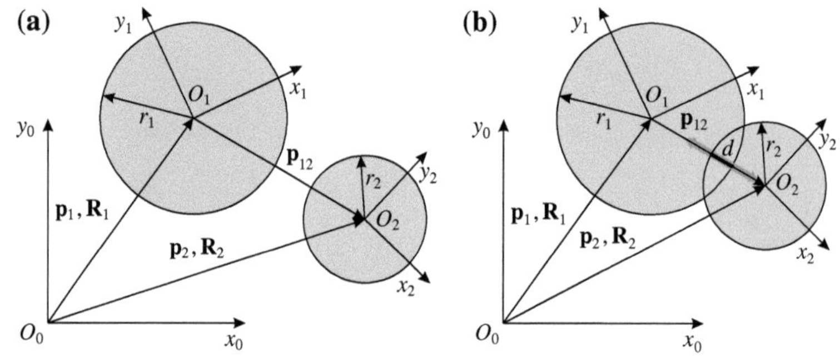

Fig. 3.12 Collision between two *spheres* (simplified view with a collision between two *circles*); the *left* image shows relations before the collision while the *right* image shows relations after the collision. Thick *straight arrows* indicate force directions

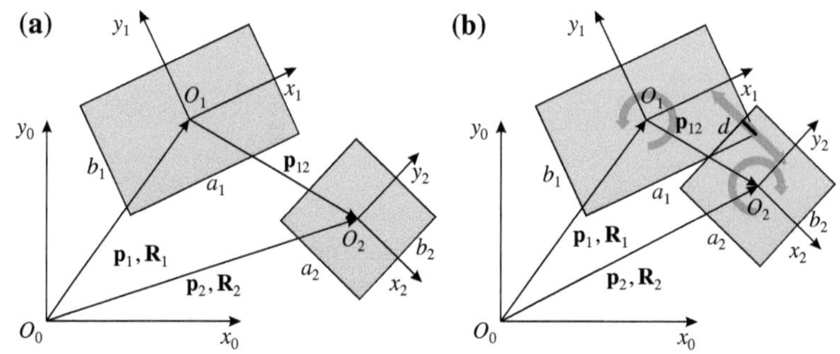

Fig. 3.13 Collision between two blocks (simplified view with a collision between two *rectangles*); the *left* image shows relations before the collision while the *right* image shows relations after the collision. Thick *straight arrows* indicate force directions, while thick *circular arrows* indicate torques acting on the blocks

the two spheres collided and the total deformation of both spheres equals

$$d = \begin{cases} 0 & \text{for } \|\mathbf{p}_{12}\| > r_1 + r_2 \\ r_1 + r_2 - \|\mathbf{p}_{12}\| & \text{for } \|\mathbf{p}_{12}\| < r_1 + r_2. \end{cases} \tag{3.29}$$

Deformation of an individual sphere can be computed based on the stiffness values (k_1, k_2) of both bodies (for example $d_1 = k_2/(k_1 + k_2)d$). In the case of frictionless collision, the reaction force direction is determined along the vector $\mathbf{p}_{12}$.

Analysis of collision between two blocks is much more complex than collision detection between two spheres. Relations are shown in Fig. 3.13.

The analysis is based on the following observation. Two convex polyhedrons are separated and do not intersect if they can be separated by a plane parallel to one of the surfaces of the two polyhedrons or with a plane that includes one of the edges of both polyhedrons.

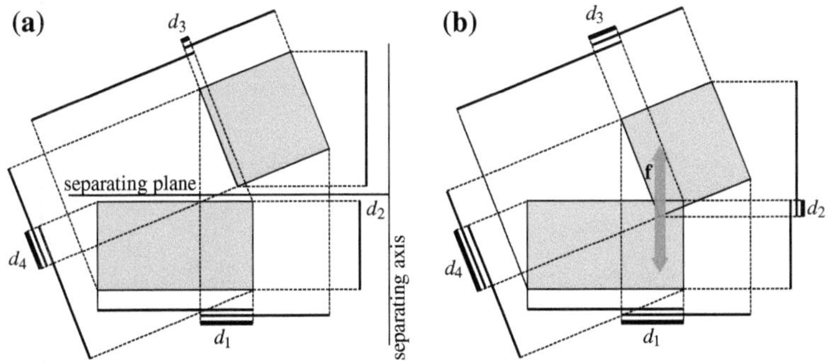

Fig. 3.14 Collision between two blocks (simplified view with a collision between two *rectangles*)—separating axis and separating plane are indicated; the *left* image shows relations before the collision while the *right* image shows relations after the collision

The existence of such a plane can be determined by projections of polyhedrons on axes that are perpendicular to the previously mentioned planes. Two convex polyhedrons are separated if there exists an axis on which their projections are separated. Such an axis is called a *separating axis*. If such an axis cannot be found, the two polyhedrons intersect.

Figure 3.14 shows collision detection between two blocks, simplified as two rectangles in a plane. Bodies are colored in grey. Around the two bodies, their projections on possible separating axes are shown. Overlap of projections may also indicate overlap of two bodies: in case (a) projections d_1, d_3 and d_4 overlap, but there is no overlap of projection d_2 (on the vertical axis). Projection d_2 defines the separating axis. In case (b) the two bodies intersect, so all their projections also overlap. In case (a) a separating plane can be found, which does not exist in case (b). It is thus possible to conclude that in case (b) the two blocks intersect and are in contact. Thus, collision occurred.

As an additional result of the analyzed collision detection algorithm, the overlap between the two blocks can be estimated. Once we compute the penetration of one body into the other, the reaction forces on the two bodies can be computed. Force direction can be determined with the direction of the separating axis with the smallest overlap. In the case shown in Fig. 3.14, projection d_2 results in the smallest overlap. Force direction is thus aligned with the vertical separating axis parallel to d_2. Since a force vector in general does not pass through the block's center of mass, the reaction force causes an additional torque that tries to rotate the block around its center of mass.

Figures 3.15 and 3.16 show relations during a collision between a block and a sphere. In this case, it is also possible to compute collisions between the two bodies based on the knowledge obtained in the previous paragraphs. As in the case of collisions between two blocks, it is necessary to compute separating planes. If a separating plane does not exist, the two bodies intersect. The separating axis with the smallest overlap can serve as an estimation of reaction force direction while the overlap itself determines the reaction force amplitude.

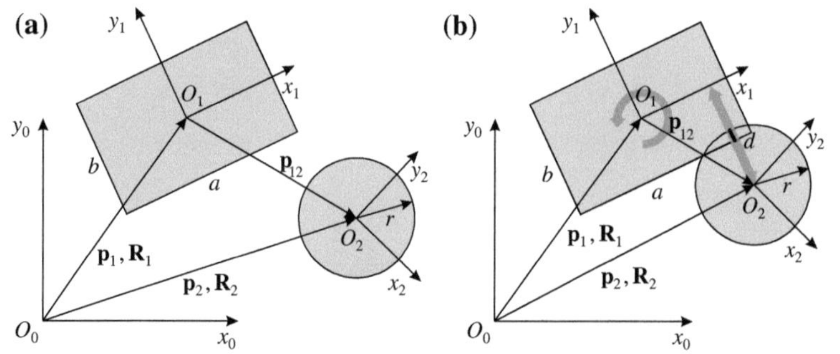

Fig. 3.15 Collision between a block and a *sphere* (simplified view with a collision between a *rectangle* and a *circle*); *left* image shows relations before the collision while the *right* image shows relations after the collision. Thick *straight arrows* indicate force directions while thick *circular arrow* indicates torque acting on the block

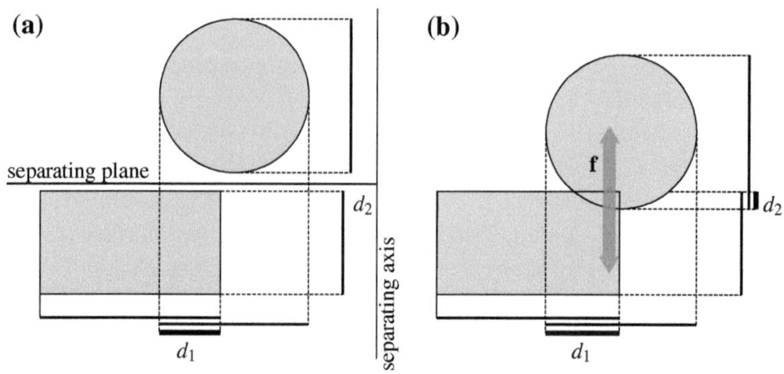

Fig. 3.16 Collision between a block and a *sphere* (simplified view with a collision between a *rectangle* and a *circle*)—separating axis and separating plane are indicated; *left* image shows relations before the collision while the *right* image shows relations after the collision

Finally, we also have to consider the problem of collisions between more complex bodies. In such cases, collision detection becomes computationally more demanding. However, it can be simplified by the use of bounding volumes as shown in Fig. 3.17. The method requires that the body being involved in collision detection is embedded into the smallest possible bounding volume. The bounding volume can take the shape of a sphere, a block or a more complex geometry such as a capsule. If a sphere is used, the bounding volume is called a *BS—bounding sphere*. The sphere is the simplest geometry of a bounding volume and enables the easiest collision detection that does not take into account the body orientation.

The use of a bounding box enables two different approaches. Method *AABB—axis aligned bounding box* assumes that box axes are always aligned with the axes of the global coordinate frame, regardless of the actual body orientation. Therefore, it becomes necessary to adjust the size of the bounding box during rotation of the body in space (middle view on Fig. 3.17). Method *OBB—oriented bounding box*

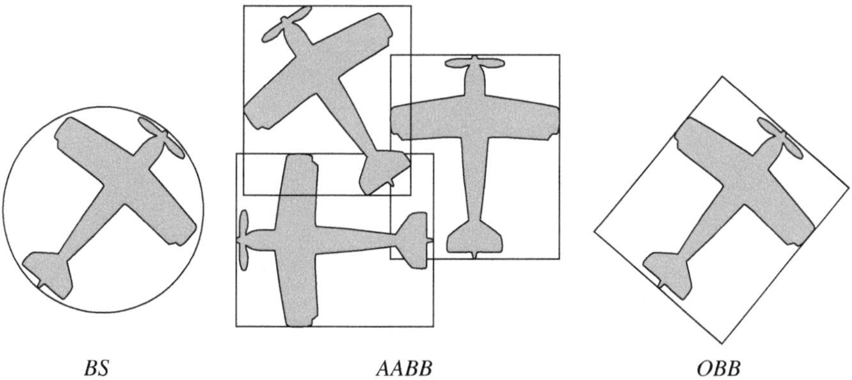

Fig. 3.17 Simplification of collision detection between complex bodies with the use of bounding volumes

assumes that the body is embedded into the smallest possible bounding box that rotates together with the rotation of the body. In this case, the bounding volume does not need to be adjusted during the rotation of the body. At the same time, *OBB* usually guarantees the most optimal representation of a body with a simplified bounding volume.

During the computation of collisions between bodies, the original (complex) geometry is replaced by a simplified geometry defined by a bounding volume. Collision detection between complex shapes can thus be translated into one of the methods addressed previously in this chapter.

The use of bounding volumes for collision detection allows only approximation of true collisions between bodies. If a simple bounding volume does not give satisfactory results, the body can be split into smaller parts and each of these parts can be embedded into its own bounding volume. The use of multiple oriented bounding boxes for a representation of a virtual object is shown in Fig. 3.18. Such representation enables a more detailed approximation of the underlying object geometry.

Fig. 3.18 Simplification of collision detection using multiple oriented bounding boxes

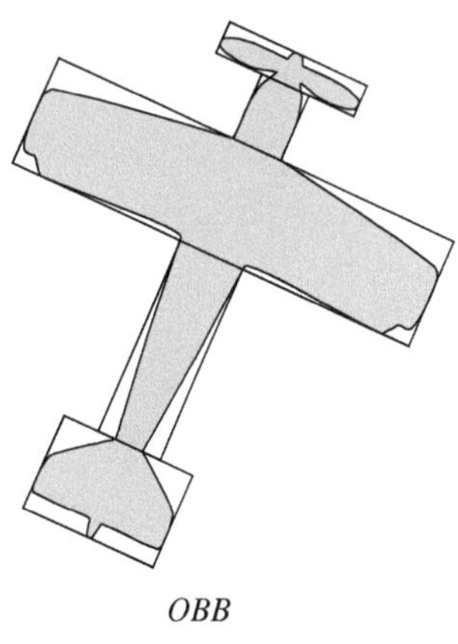

3.6 Computation of Body Motion

After an introduction of the basic equations of motion, we can now implement an algorithm for computation of body motion in a computer-generated environment. A block scheme representing the basic concept is shown in Fig. 3.19. It is possible to note that the algorithm consists of a closed loop that needs to be computed within a single computational time step of the rendering loop.

Since the loop in Fig. 3.19 is closed, there is no natural entry point. Therefore, we will follow labels in the order from ① to ④ We assume the force to be the cause of body motion. Point ① represents the sum of all forces and torques (user, virtual actuators, medium, collision, field) acting on the body. Forces resulting from possible collisions with other bodies can be computed based on the initial pose of the body, determined with $\mathbf{p}_0$ and q_0. Interaction force with the medium can be computed based on the initial body velocity (that can be determined from the initial body linear and angular momenta) while interaction force with the user can be determined with the initial user's activity. A force field is usually time-independent (for example, a gravity field).

Based on Eqs. (3.19) and (3.20), body linear and angular momenta can be computed by the time integral of forces and torques. Since a simulation loop runs on a computer with a discrete sampling time Δt, Eqs. (3.19) and (3.20) need to be discretized as well

$$\begin{aligned} \mathbf{G}_{k+1} &= \mathbf{G}_k + \mathbf{F}_k \Delta t \\ \boldsymbol{\Gamma}_{k+1} &= \boldsymbol{\Gamma}_k + \boldsymbol{\tau}_k \Delta t, \end{aligned} \tag{3.30}$$

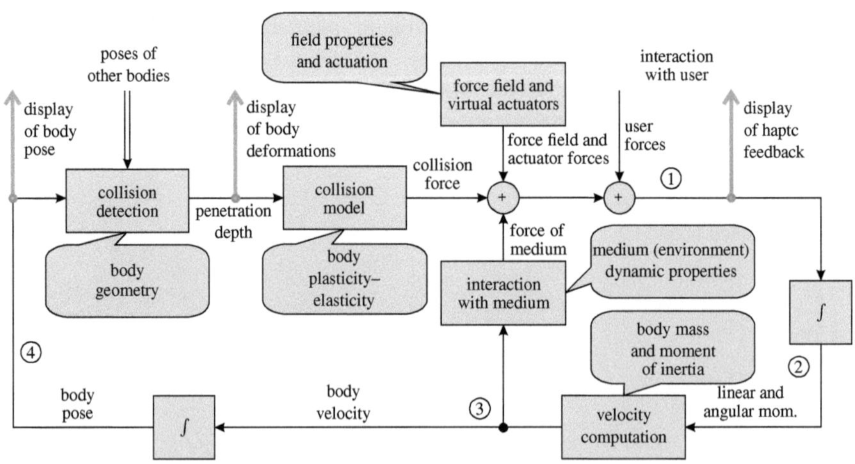

Fig. 3.19 Algorithm for computation of body motion as a result of interactions with other bodies, with the user and with the medium. The force represents a generalized quantity that includes also torques

where $\mathbf{G}_k$ and $\boldsymbol{\Gamma}_k$ represent body linear and angular momenta at discrete time interval k, respectively. Initial linear and angular momenta are $\mathbf{G}_0$ and $\boldsymbol{\Gamma}_0$. The result of integration is marked with label ② in Fig. 3.19.

From the linear and angular momenta computed from (3.13) and (3.14) and by taking into account body inertial properties (mass and moment of inertia), body translational velocity $\mathbf{v}_{k+1}$ and rotational velocity $\boldsymbol{\omega}_{k+1}$ can be computed for time instant $k+1$ (label ③)

$$\begin{aligned} \mathbf{v}_{k+1} &= \frac{1}{M}\mathbf{G}_{k+1} \\ \boldsymbol{\omega}_{k+1} &= \mathbf{I}_k^{-1}\boldsymbol{\Gamma}_{k+1}, \end{aligned} \tag{3.31}$$

where $\mathbf{I}_k$ represents the body inertia in relation to the global coordinate frame at time instant k.

The new object pose can be computed based on Eqs. (3.2) and (3.6). These equations are numerically integrated

$$\begin{aligned} \mathbf{p}_{k+1} &= \mathbf{p}_k + \mathbf{v}_{k+1}\Delta t \\ q_{k+1} &= q_k + \frac{1}{2}\Delta t\, \tilde{\omega}_{k+1} \otimes q_k, \end{aligned} \tag{3.32}$$

where $\mathbf{p}_k$ and q_k are the body position and orientation at time interval k, while initial position and orientation are determined with $\mathbf{p}_0$ and q_0. The new body pose is now computed (label ④) as a consequence of interactions with other bodies, with the user, with the medium and due to the effects of virtual actuators. The loop for computation of body pose continues in point ① after all interaction forces are computed. The loop presented in Fig. 3.19 needs to be implemented for all dynamic bodies in the virtual environment.

Chapter 4
Tracking the User and Environment

Abstract Virtual reality allows different methods of interaction and communication between the user and virtual world. An important aspect of the virtual reality system in this regard is the tracking of the user's pose and actions. In this chapter methods and technologies that define inputs to a virtual reality system are presented. These include pose sensors (mechanical, ultrasonic, optical, electromagnetic, and inertial) as well as force and torque sensors. The chapter concludes with a summary of motion tracking concepts and physical input devices.

Virtual reality allows different methods of communication between the user and virtual world. If we want to create a feeling of presence in the synthetic environment, we need equipment that can track the user's position and actions [1]. Information about the user's actions allows the system to show the virtual environment from the user's perspective, a basic requirement for the induction of physical virtual presence. At the same time, inputs provided by the user allow interaction with the virtual world. The user's interaction with the virtual world via a virtual reality system allows two-way exchange of information via input/output devices.

The user's movement and actions can be tracked using either active methods (triggered by the user), where the user transmits information to the virtual reality system, or using passive methods (triggered by the system without the user's cooperation), which sense the user's movement and inform the computer about the user's position and point of gaze. Active methods include spoken instructions as well as controllers such as joysticks, keyboards, wheels or gamepads. Passive tracking methods are summarized in Fig. 4.1 [2].

In addition to tracking the user, it is also sometimes necessary to track the environment so that information from the real world can be combined with the virtual world. The real world is usually observed with sensors that are not directly connected to the user. Inputs from the real world are frequently used to create parts of the virtual world in real time.

M. Mihelj et al., *Virtual Reality Technology and Applications*,
Intelligent Systems, Control and Automation: Science and Engineering 68,
DOI: 10.1007/978-94-007-6910-6_4,

Tracking methods in virtual reality are primarily used for:

(a) *View control*: Motion tracking devices allow control of the position and orientation of the camera used to create an image of the virtual environment for head-mounted or projection displays. This allows the virtual environment to be displayed from the user's perspective. As the user moves around the room, the view of the virtual environment changes accordingly.
(b) *Navigation*: Motion tracking devices allow the user to navigate inside the virtual environment.
(c) *Object selection and manipulation*: Tracking the user's hands and the objects held by the user allows objects to be selected and manipulated in the virtual environment.
(d) *Tool tracking*: Tracking a tool (e.g. one held in the user's hand) allows synchronization between the position of virtual tools in the virtual world and tools manipulated by the user in the real world.
(e) *Avatar animation*: User motion tracking methods allow realistic movement of avatars in the virtual environment.

Figure 4.1 shows that motion tracking methods can be based on different physical principles. There is thus no simple method that would allow perfect motion tracking. An ideal tracking device would need to satisfy several often contradictory demands that can be summarized as follows [3]:

(a) *Small*: The device should be the size of a chip or even smaller.
(b) *Self-contained*: Should not require additional elements placed on the human or environment.
(c) *Complete*: Should allow motion tracking in all six degrees of freedom (three position, three orientation).
(d) *Accurate*: Should have a sufficiently high measurement resolution (<1 mm and <0.1°).

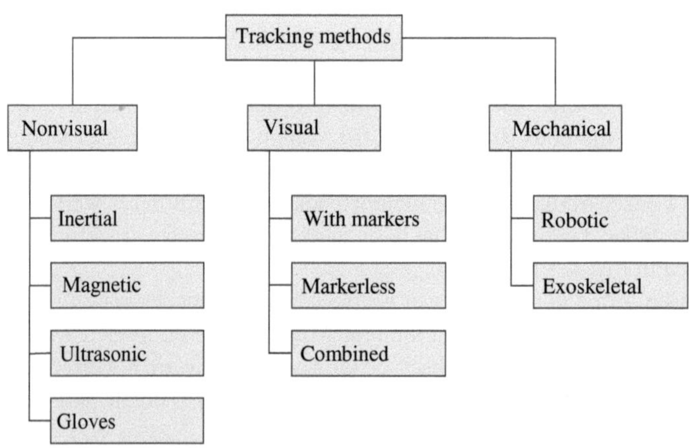

Fig. 4.1 Different user motion tracking methods

(e) *Fast*: Should allow sampling frequencies on the order of 1 kHz, regardless of the number of included devices.
(f) *Insensitive to occlusion*: The device should not need a direct view of the sensor.
(g) *Robust*: Should be robust with regard to external influences (temperature, moisture, magnetic field, radiofrequency noise).
(h) *Unlimited working area*: Should allow a target to be tracked regardless of its speed and distance.
(i) *Wireless*: Should work without any wires, only with battery power and wireless connection to a computer.
(j) *Cheap*.

4.1 Pose Sensor

A pose sensor is a device that allows both the position and orientation to be tracked. It is perhaps the most important measurement device in a virtual reality system, as it allows the system to detect the position and orientation of the user in the virtual world. The device also sets the limitations of virtual reality. Pose tracking methods are usually based on electromagnetic, mechanical, optical, videometric, ultrasonic or inertial principles. Each of these has its own advantages and disadvantages. Disadvantages are mainly due to limitations of the specific physical medium and limitations of specific devices or signal processing techniques.

4.1.1 Mechanical Principle

Conceptually the simplest principle of user motion tracking, the mechanical principle assumes a direct physical connection between the user and the measurement device. A typical approach uses an articulated mechanism with two or more segments connected by joints whose angles can be measured. The mechanism thus has multiple degrees of freedom and is connected to the user. The device follows the user's motions. The mechanism can also be equipped with weight compensation for higher effectiveness.

Let's illustrate the mechanical measurement principle with a simple example shown in Fig. 4.2. Let's assume that the user is touching the end of the mechanism shown in the figure. The user's pose is thus defined through the pose of coordinate system $[x_3, y_3, z_3]$. Our goal is the geometric model of the mechanical device. This model describes the pose of the coordinate system at the end of the device with regard to the base coordinate system and can be obtained by consecutively multiplying (postmultiplication) homogeneous transformation matrices. However, in this case the model is relatively simple and can be calculated directly from the relations in Fig. 4.2b, which gives a bird's eye view of the mechanism.

Since the mechanism has only two degrees of freedom, the motion of the endpoint is limited to the xy plane. The height is constant and determined by the length of

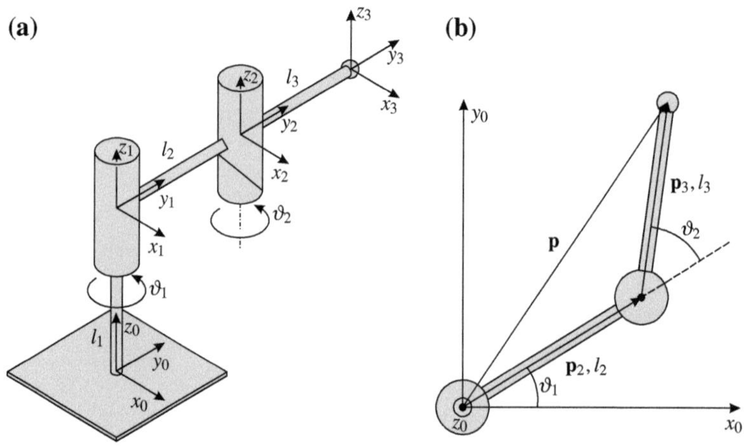

Fig. 4.2 Example of a mechanism with two degrees of freedom (**a**). Kinematic model of the mechanism (**b**)

the first segment l_1. Along this segment, we define the vector $\mathbf{p}_1 = \begin{bmatrix} 0 \; 0 \; l_1 \end{bmatrix}^T$. The rotation axis for the first joint is the vertical axis z_0, which points out of the page in Fig. 4.2b. We define vector $\mathbf{p}_2$ in the direction of the second segment. This gives

$$\mathbf{p}_2 = l_2 \begin{bmatrix} \cos \vartheta_1 \\ \sin \vartheta_1 \\ 0 \end{bmatrix}. \tag{4.1}$$

Vector $\mathbf{p}_3$ goes along the third segment. Its components can be determined from Fig. 4.2b

$$\mathbf{p}_3 = l_3 \begin{bmatrix} \cos(\vartheta_1 + \vartheta_2) \\ \sin(\vartheta_1 + \vartheta_2) \\ 0 \end{bmatrix}. \tag{4.2}$$

We also define vector $\mathbf{p}$, which goes from the origin of the coordinate system (x_0, y_0, z_0) to the end of the robot

$$\mathbf{p} = \mathbf{p}_1 + \mathbf{p}_2 + \mathbf{p}_3. \tag{4.3}$$

Vector $\mathbf{p}$ represents the position of the end of the manipulator

$$\mathbf{p} = \begin{bmatrix} x \\ y \\ z \end{bmatrix} = \begin{bmatrix} l_2 \cos \vartheta_1 + l_3 \cos(\vartheta_1 + \vartheta_2) \\ l_2 \sin \vartheta_1 + l_3 \sin(\vartheta_1 + \vartheta_2) \\ l_1 \end{bmatrix}. \tag{4.4}$$

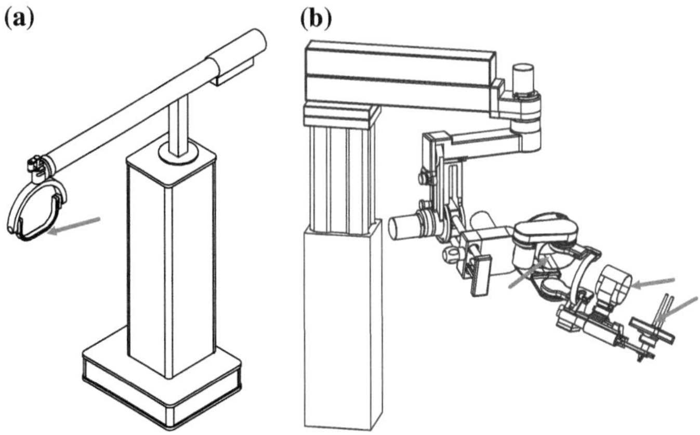

Fig. 4.3 Examples of two mechanical devices. **a** The robot in the figure is in contact with the user at only one point at the end of the device (*arrow*). **b** Shows an exoskeletal robot where interaction between the human and device occurs at many points (*arrows*)

Figure 4.3 shows examples of two more mechanical devices. The robot in Fig. 4.3a is in contact with the user at one point at the end of the device (arrow). In this case, we only need to track the endpoint of the user's limb. Figure 4.3b shows an exoskeletal robot where interaction between the robot and user occurs at many points (arrows). The exoskeletal mechanism allows all segments of the limb to be tracked, not only its endpoint.

Finally, a few strengths and weaknesses of the mechanical tracking principle should be mentioned. It allows position and orientation to be tracked with a high accuracy that mostly depends on the specifics of the optical encoders used to measure joint angles. Sampling frequencies over 1 kHz are attainable, and delays are small. Weaknesses include high complexity, high price and motion constraints introduced by the measurement mechanism.

4.1.2 Ultrasonic Principle

The ultrasonic principle is based on high-frequency sound that allows measurement of the distance between an emitter (speaker) and receiver (microphone). In dry air with a temperature of 20 °C, the speed of sound is $c = 343.2$ m/s. The distance between the emitter and receiver can thus be calculated as

$$l = ct_{us}, \tag{4.5}$$

where t_{us} is the travel time of the ultrasonic pulse from the emitter to the receiver. Three noncollinear receivers are required to determine the position of a point in space.

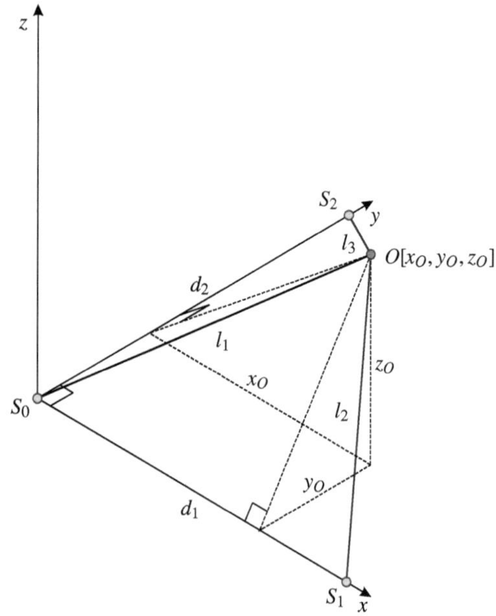

Fig. 4.4 Triangulation used to determine the position of a point based on the ultrasonic principle

The situation is illustrated in Fig. 4.4 [4]. To simplify calculations, receivers S_0, S_1 and S_2 are located at fixed points along the axes of a Cartesian coordinate system. The ultrasound emitter is placed at measurement point $O[x_O, y_O, z_O]$. Equation 4.5 allows us to calculate distances between the emitter and receivers l_1, l_2 and l_3. If we know the distances between receivers d_1 and d_2, it is possible to determine the coordinates of the emitter in the coordinate system $[x, y, z]$ as

$$\begin{aligned} x_O &= \frac{l_1^2 + d_1^2 - l_2^2}{2d_1} \\ y_O &= \frac{l_1^2 + d_2^2 - l_3^2}{2d_2} \\ z_O &= \sqrt{l_1^2 - x_O^2 - y_O^2}. \end{aligned} \tag{4.6}$$

If we wish to determine the orientation of an object in space, we must know the positions of at least three noncollinear points on the tracked object. The number of emitters must thus be increased to at least three. The situation is shown in Fig. 4.5. Emitters O_0, O_1 and O_2 are attached to the tracked object. Three receivers S_0, S_1 and S_2 measure the ultrasonic pulses sequentially emitted by the emitters. The spatial coordinates of the three emitters can be calculated based on nine measurements of ultrasound travel time (three emitters and three receivers), thus allowing the axes of the object's coordinate system to also be determined.

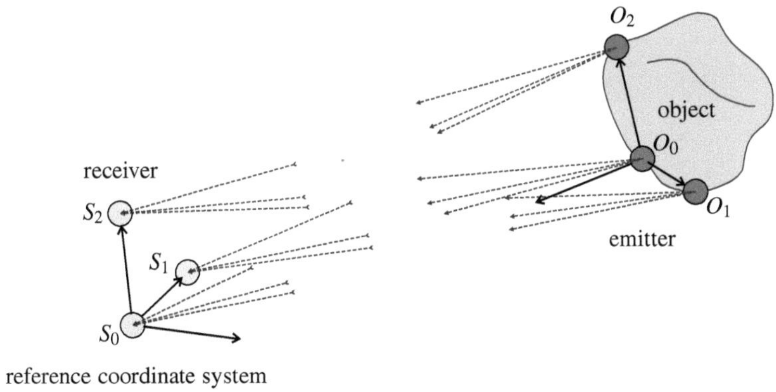

Fig. 4.5 Determining the pose of an object based on the ultrasonic principle

Finally, a few strengths and weaknesses of the ultrasonic tracking principle should be mentioned. The greatest weaknesses are caused by measurements of the ultrasound travel time from the emitter to the receiver. The speed of sound depends on temperature, pressure and humidity as well as any barriers in the ultrasound's path. All of these effects can decrease the accuracy of distance measurements and thus also the accuracy of the tracked object's pose. An additional weakness of ultrasonic measurements is the relatively low speed of sound in air, which limits the highest attainable sampling frequency to a few dozen Hz and causes nonnegligible measurement delays (up to a few dozen ms). The strengths of the principle include simple and cheap measurement technology as well as relatively small sensors.

4.1.3 Optical Principle

Optical trackers use visual information to track the user's motion. Measurements can be performed using videocameras or special cameras with active or passive markers.

The task of computer vision is to recognize the geometry of the scene or the movement of the user from a digital image. Let's start the explanation with the simplest example: a single point (Fig. 4.6) that can also represent a marker attached to the person. We want to determine the relationship between the point's coordinates in a two-dimensional image and the point's coordinates in the three-dimensional real world. The basic equations of optics tell us that the position of the point in the image plane depends on the position of the same point in three-dimensional space. Our task is to find the geometric relationship between the coordinates of point $P[x_c, y_c, z_c]$ in space and the coordinates of the same point $p[u, v]$ in the image.

Since the opening in the lens through which light reaches the image plane is small compared to the size of the observed objects, the camera lens can be replaced with a pinhole in the mathematical model. In a perspective transformation, all points

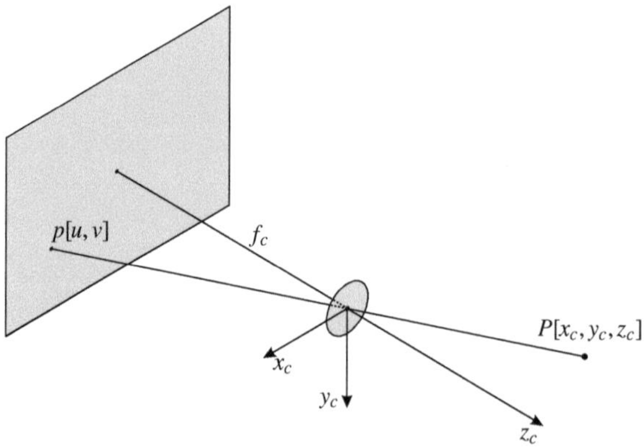

Fig. 4.6 Perspective projection

are mapped to the same plane via lines that intersect in a point called the center of projection. When a real camera is replaced by a camera with a pinhole, the center of projection is in the center of the lens.

A coordinate system must be attached to the camera. This allows the pose of the camera to always be described through the pose of the selected coordinate system. Axis z_c of the camera's coordinate system points along the optical axis while the origin of the coordinate system is placed in the center of projection. We choose a right-handed coordinate system where axis x_c is parallel to the rows of the image while axis y_c is parallel to its columns.

In a camera, the image plane is located behind the center of projection. The distance f_c between the image and the center of projection is called the focal length. In the coordinate system of the camera, the focal length has a negative value since the image plane is in the negative part of axis z_c. For our model, it is more convenient to use the equivalent image plane on the positive side of axis z_c (Fig. 4.7).

The equivalent and real image planes are symmetric with regard to plane $[x_c, y_c]$ of the camera coordinate system. The geometric properties of objects in both planes are completely equivalent and differ only in their signs.

From now on, we will refer to the equivalent image plane simply as the image plane. The image plane can be considered a rigid body, so a coordinate system can also be attached to it. The coordinate origin is placed in the intersection of the optical axis and the image plane. Axes x_s and y_s should be parallel to axes x_c and y_c of the camera coordinate system.

The camera thus has two coordinate systems: the camera coordinate system and the image plane system. If point P is expressed in the camera coordinate system and p represents the projection of point P onto the image plane, we are interested in the relation between the coordinates of point P and the coordinates of point p.

Let's say that point P lies on the $[y_c, z_c]$ plane of the camera coordinate system. Its coordinates are then

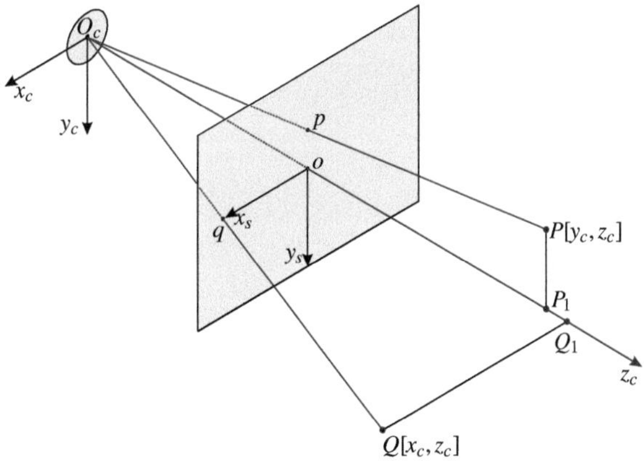

Fig. 4.7 Equivalent image plane

$$P = \begin{bmatrix} 0 \\ y_c \\ z_c \end{bmatrix}. \tag{4.7}$$

Projection p then falls onto the y_s axis of the image coordinate system

$$p = \begin{bmatrix} 0 \\ y_s \end{bmatrix}. \tag{4.8}$$

Due to the similarity of triangles PP_1O_c and poO_c, we may write

$$\frac{y_c}{y_s} = \frac{z_c}{f_c}$$

or

$$y_s = f_c \frac{y_c}{z_c}. \tag{4.9}$$

Let's also take point Q, which lies in the $[x_c, z_c]$ plane of the camera coordinate system. In a perspective projection of point Q, its image q falls on the x_s axis of the image coordinate system. Due to the similarity of triangles QQ_1O_c and qoO_c, we write

$$\frac{x_c}{x_s} = \frac{z_c}{f_c}$$

and

$$x_s = f_c \frac{x_c}{z_c}. \tag{4.10}$$

We have thus obtained the relation between the coordinates of point $P = [x_c, y_c, z_c]^T$, expressed in the camera space, and the point $p = [x_s, y_s]^T$, expressed in the image space. The above equations represent the mathematical description of the perspective projection from $3D$-space to a $2D$-space. They can be written in matrix form

$$\lambda \begin{bmatrix} x_s \\ y_s \\ 1 \end{bmatrix} = \begin{bmatrix} f_c & 0 & 0 & 0 \\ 0 & f_c & 0 & 0 \\ 0 & 0 & 1 & 0 \end{bmatrix} \begin{bmatrix} x_c \\ y_c \\ z_c \\ 1 \end{bmatrix}. \tag{4.11}$$

In Eq. (4.11), λ is the scaling factor, $[x_s, y_s, 1]^T$ are the projected coordinates of the point in the image coordinate system, and $[x_c, y_c, z_c, 1]^T$ are the coordinates of the original point in the camera coordinate system. We also define the perspective projection matrix

$$\boldsymbol{\Pi} = \begin{bmatrix} f_c & 0 & 0 & 0 \\ 0 & f_c & 0 & 0 \\ 0 & 0 & 1 & 0 \end{bmatrix}. \tag{4.12}$$

It is easy to see from the Eq. (4.11) that coordinates $[x_s, y_s, \lambda]^T$ can be unambiguously determined when $[x_c, y_c, z_c, 1]^T$ is known. However, it is not possible to determine the coordinates $[x_c, y_c, z_c]^T$ in the camera coordinate system if only the coordinates $[x_s, y_s]^T$ in the image coordinate system are known and the scaling factor λ is unknown. The above matrix equation represents a direct projection while the calculation of $[x_c, y_c, z_c]^T$ from $[x_s, y_s]^T$ is called the inverse projection. If only one camera is used and we have no foreknowledge about the size of the objects in the scene, it is impossible to find an unambiguous solution to the inverse problem.

Next we examine the inverse projection. As previously stated, the direct relationship is (4.11)

$$\lambda \begin{bmatrix} x_s \\ y_s \\ 1 \end{bmatrix} = \boldsymbol{\Pi} \begin{bmatrix} x_c \\ y_c \\ z_c \\ 1 \end{bmatrix}. \tag{4.13}$$

We now need to determine the coordinates $[x_c, y_c, z_c]$ if the point's coordinates in the image and the matrix $\boldsymbol{\Pi}$ are known. The scaling factor λ is unknown. We thus need four equations to solve the problem since there are four unknown variables λ, x_c, y_c and z_c. For one point, we have three equations and four unknown variables. Let's try with three points (Fig. 4.8).

We now have three points (A, B, C) and know the distances between them. Their coordinates in the camera coordinate system are

$$\left\{(x_{c,i}, y_{c,i}, z_{c,i}), \quad i = 1, 2, 3\right\}.$$

The coordinates of the corresponding points in the image are

$$\left\{(x_{s,i}, y_{s,i}), \quad i = 1, 2, 3\right\}.$$

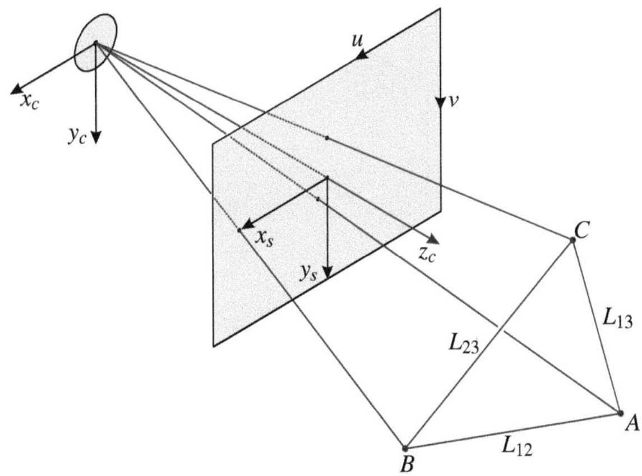

Fig. 4.8 Example of projecting three points

We then write the direct projection mapping in the following form

$$\lambda_i \begin{bmatrix} x_{s,i} \\ y_{s,i} \\ 1 \end{bmatrix} = \boldsymbol{\Pi} \begin{bmatrix} x_{c,i} \\ y_{c,i} \\ z_{c,i} \\ 1 \end{bmatrix}. \tag{4.14}$$

The system (4.14) contains 12 unknown variables and 9 equations. The solution can be found only with three additional equations. These can be obtained from the size of the triangle created by points A, B and C. The lengths of sides AB, BC and CA can be labeled with distances L_{12}, L_{23} and L_{31}

$$\begin{aligned} L_{12}^2 &= (x_{c,1} - x_{c,2})^2 + (y_{c,1} - y_{c,2})^2 + (z_{c,1} - z_{c,2})^2 \\ L_{23}^2 &= (x_{c,2} - x_{c,3})^2 + (y_{c,2} - y_{c,3})^2 + (z_{c,2} - z_{c,3})^2 \\ L_{31}^2 &= (x_{c,3} - x_{c,1})^2 + (y_{c,3} - y_{c,1})^2 + (z_{c,3} - z_{c,1})^2. \end{aligned} \tag{4.15}$$

We now have twelve equations for twelve unknown variables. A solution to the problem thus exists. Unfortunately, the last three equations are nonlinear and must be solved numerically with special software. This solving method is called an inverse projection mapping based on a model.

Since the model of the observed object is usually not available or the object changes with time (e.g. a walking human), other solutions to the inverse projection mapping problem need to be found. One possible solution is the use of stereo vision: sensing based on two cameras. The principle is similar to human visual perception where the images seen by the left and right eyes differ slightly due to parallax and the brain uses the differences between images to determine the distance to the observed object.

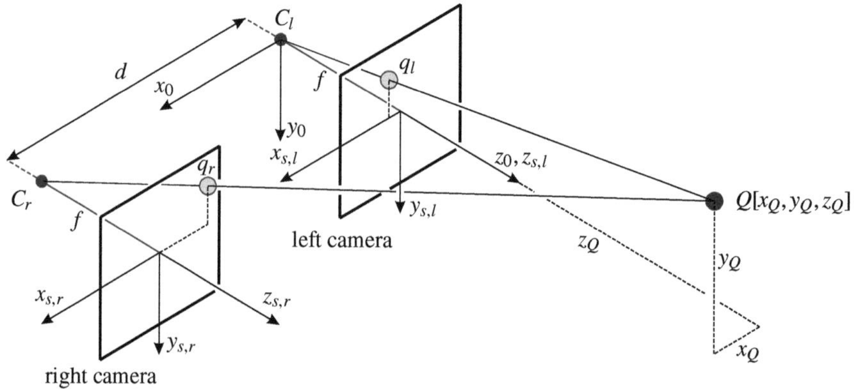

Fig. 4.9 Stereo view of point O using two parallel cameras

The principle of using two parallel cameras to observe point Q is shown in Fig. 4.9. Point Q is projected onto the image plane of the left and right cameras. The left camera's image plane contains projection q_l with coordinates $x_{s,l}$ and $y_{s,l}$ while the right camera's image plane contains projection q_r with coordinates $x_{s,r}$ and $y_{s,r}$. The axes of the reference coordinate system $[x_0, y_0, z_0]$ have the same directions as the left camera's coordinate system.

Figure 4.10a shows the top view while Fig. 4.10b shows the side view of the situation in Fig. 4.9. These views will help us calculate the coordinates of point Q. From the geometry in Fig. 4.10a we can extract the following relations (distances x_Q, y_Q and z_Q are with regard to coordinate system $[x_0, y_0, z_0]$)

$$\begin{aligned} \frac{z_Q}{f} &= \frac{x_Q}{x_{s,l}} \\ \frac{z_Q}{f} &= \frac{x_Q - d}{x_{s,r}} \end{aligned} \tag{4.16}$$

From the above equation in (4.16) we express

$$x_Q = \frac{x_{s,l}}{f} z_Q \tag{4.17}$$

and insert into the second equation to get

$$\frac{x_{s,l} z_Q}{x_{s,r} f} - \frac{z_Q}{f} = \frac{d}{x_{s,r}}. \tag{4.18}$$

We can then determine the distance z_Q to point Q as

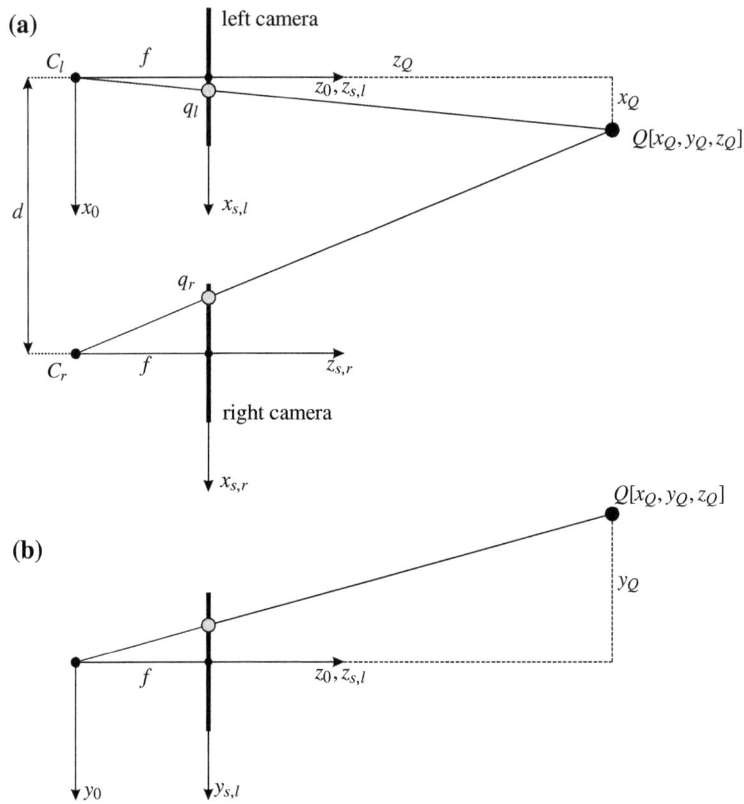

Fig. 4.10 Projections of point O on the planes of the *left* and *right cameras*. **a** Shows a view of both cameras from *above* while **b** shows a side view of the cameras

$$z_Q = \frac{fd}{x_{s,l} - x_{s,r}}. \tag{4.19}$$

The distance x_Q can be determined from Eq. (4.17). To determine distance y_Q we take into account the situation in Fig. 4.10b. From the geometry we can extract relation

$$\frac{z_Q}{f} = \frac{y_Q}{y_{s,l}}, \tag{4.20}$$

allowing us to calculate the remaining coordinate

$$y_Q = \frac{y_{s,l}}{f} z_Q. \tag{4.21}$$

Using two cameras thus allows the position (and orientation) of an object in space to be determined without an accurate model of the object. This naturally greatly

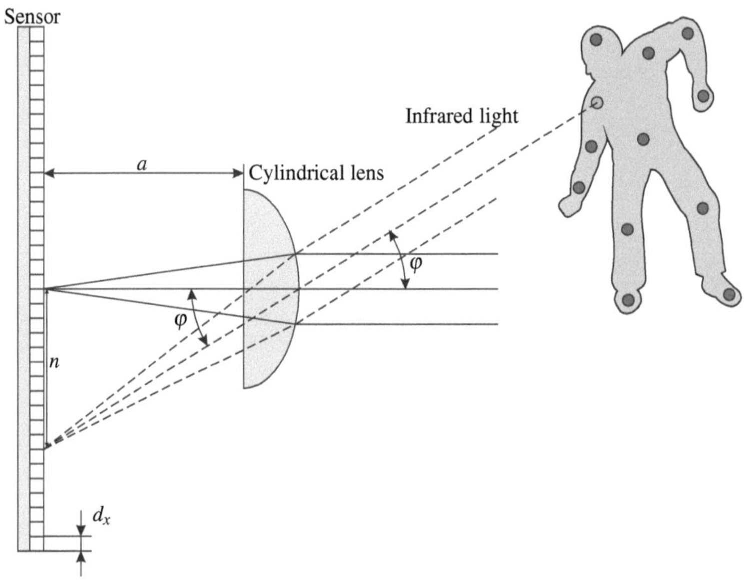

Fig. 4.11 CCD line sensor

simplifies tracking of various objects, but requires a well-calibrated stereo vision system.

To simplify human motion tracking, special markers (usually reflective spherical objects) are frequently attached to characteristic points on the body and tracked with optical systems. This avoids the need to track every point of the body, allowing more accurate and faster calculation of spatial coordinates. Such markers can be used together with a stereo camera system such as the one analyzed above. Frequently, the camera system can even be simplified with a line of optical sensors that replace the camera image plane. An example of a line sensor together with an infrared (IR) light emitter (marker) is shown in Fig. 4.11. The light from the IR-marker is focused onto the line sensor by a cylindrical lens. The point on the line sensor where the marker light is focused depends on the spatial position of the marker. From the distance between the lens and the sensor a, the position of the illuminated pixel n, and its width d_x, it is possible to calculate the incidence angle of the incoming light

$$\varphi = \arctan \frac{nd_x}{a}. \tag{4.22}$$

A better resolution of position measurements can be obtained with a line sensor with a smaller pixel width d_x. The position of multiple markers can be determined by triggering one infrared emitter after another.

Of course, the spatial position of the marker cannot be determined with a single line sensor. It requires at least three sensors with different orientations. An example

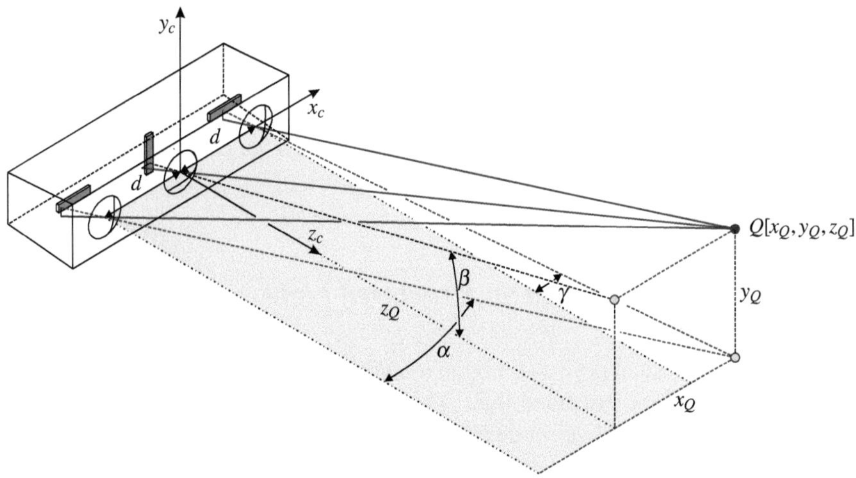

Fig. 4.12 Optical triangulation based on three optical line sensors

of a camera system with three line sensors is shown in Fig. 4.12. The position of the IR marker in the $[x_c, y_c, z_c]$ coordinate system can be calculated from the three incidence angles of the IR rays onto the three line sensors. These three angles are marked as α, β and γ. Incidence angle α is determined as

$$\tan\alpha = \frac{x_Q + d}{z_Q}, \tag{4.23}$$

incidence angle β is determined as

$$\tan\beta = \frac{y_Q}{z_Q} \tag{4.24}$$

and incidence angle γ is determined as

$$\tan\gamma = \frac{x_Q - d}{z_Q}. \tag{4.25}$$

From Eqs. (4.23) and (4.25) we can calculate the distance

$$z_Q = \frac{2d}{\tan\alpha - \tan\gamma}. \tag{4.26}$$

Equation (4.23) then allows us to determine the distance

$$x_Q = z_Q \tan\alpha - d \tag{4.27}$$

Finally, from Eq. (4.24) we determine the distance

$$y_Q = z_Q \tan \beta. \tag{4.28}$$

It is thus possible to quickly and accurately track a large number of infrared markers attached to measured users or objects.

4.1.4 Triangulation Based on Structured Light

No discussion of optical pose and motion tracking is complete without a mention of triangulation based on structured light. The principle of this technology originates from metrology and allows accurate measurement of object distances or thicknesses. The basic principle of triangulation using a laser light source is shown in Fig. 4.13. A camera senses laser light reflected from the object. It is at a distance of d_p and angle φ away from the light source. The figure illustrates that a vertical movement of the object for a distance d_x also causes the projection of the laser point on the camera to move a distance of d_y. The displacement on the camera thus allows the displacement of the object to be calculated.

Triangulation that allows object motion tracking is based on the projection of a light pattern onto the observed object. The simplest pattern consists of parallel stripes

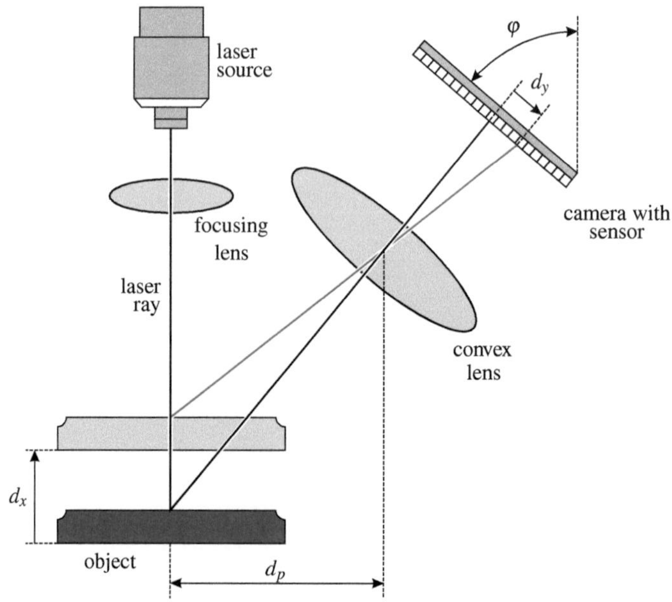

Fig. 4.13 Principle of laser-aided triangulation

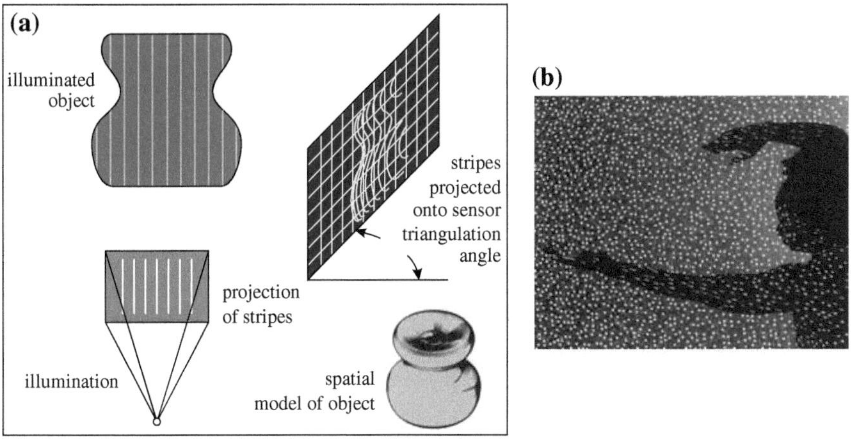

Fig. 4.14 **a** Triangulation based on structured light. **b** Projection of dot grid

of light as shown in Fig. 4.14a. From the viewpoint of the projector, the pattern consists of straight lines projected onto the object. The camera, which is displaced by a certain triangulation angle, perceives the same stripes as curves running along the surface of the object. Mathematical algorithms allow the three-dimensional shape of the object to be determined from the shape of the stripes.

The principle of structured light can also be used to sense the depth or distance of objects. An appropriate illumination pattern for such a purpose is shown in Fig. 4.14b. In the area where the object is located, the distribution of dots is different than in the background. The depth of the object also affects the pattern of dots. Image processing allows information about objects' distances to be obtained.

Additional processing of object distance data also allows tracking of pose or motion if the observed person or object is located in the camera's field of view [5]. The procedure is illustrated in Fig. 4.15. The first figure shows distance information (closer objects are darker). Depth information is used to perform segmentation as seen in the second figure. The segmented image allows the reconstruction of the person's skeleton, which gives information about segment positions relative to one another (third image). To perform this segmentation and reconstruct the skeleton, the algorithm learns from a training database of measurements that describe human kinematic properties. A detailed description of such learning, however, is beyond the scope of this work.

4.1.5 Videometric Principle

The videometric principle is actually the optical principle, except that the camera is not fixed in space but is instead attached to the object whose pose we wish to

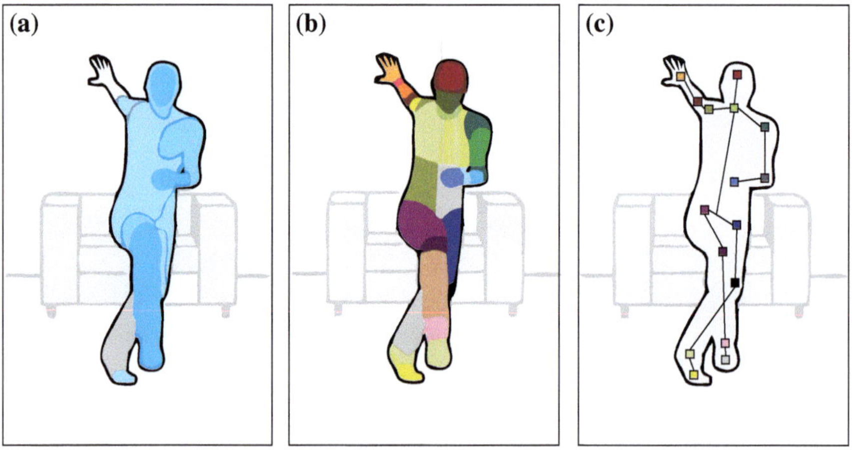

Fig. 4.15 Human skeleton reconstruction based on triangulation using structured light

determine. Markers placed around the room are necessary for the videometric principle and used to determine the pose of the object to which the camera is attached.

4.1.6 Radiofrequency Principle

The radiofrequency principle is rarely used for user motion tracking in virtual reality and is thus mentioned here only briefly. Most methods are based on measuring the travel time of a radiofrequency signal and thus require very accurate measurements of time. The principle is not suitable for measuring short distances with high accuracy, and the complexity of the measurements results in expensive equipment.

An example of the radiofrequency principle applied to user motion tracking is GPS, which is not used in virtual reality. The measurement concept is shown in Fig. 4.16. The measurement is based on measuring signal travel time from a satellite to a user on the Earth's surface. Position calculations are based on triangulation, which is done as in the ultrasonic principle and thus not repeated here.

4.1.7 Electromagnetic Principle

The principle is based on measuring the local vector of the magnetic field in the sensor's surroundings. The Earth's magnetic field can be used as a basis for measurement, but such measurements are usually not accurate enough. Thus, it is usually necessary to create an additional magnetic field especially for the measurement.

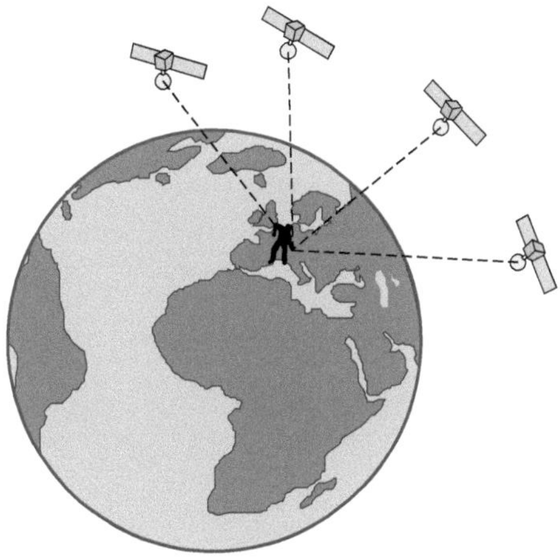

Fig. 4.16 Example of placing emitters in space and sensors on the person—Global Positioning System

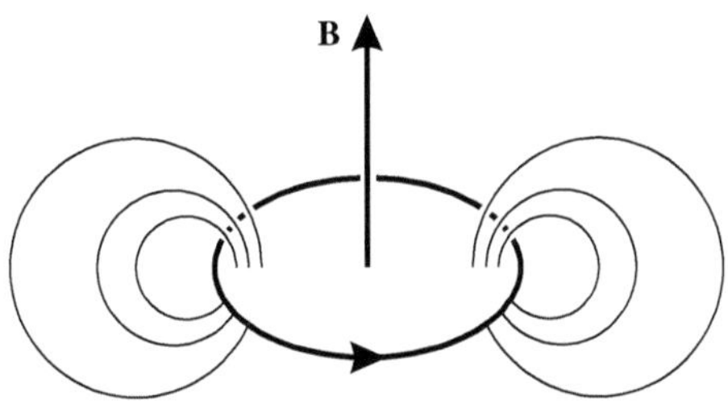

Fig. 4.17 Structure of a magnetic field in a simple coil

A simple coil stimulated with an electric current creates an electromagnetic field (Fig. 4.17). In an empty room, the field is symmetric around an axis perpendicular to the plane of the coil. The density and polarization of the magnetic field are represented by vector **B**. The density of the magnetic field is proportional to the field's strength **H**

$$\mathbf{B} = \mu_0 \mathbf{H}, \tag{4.29}$$

where μ_0 is the magnetic constant.

Without going too far into detail, let's write the coil's magnetic field density as a function of distance from the coil. Figure 4.18 shows a circular electrical loop with radius b, through which the current I flows. Taking into account $R \gg b$, the coil can be approximated as a magnetic dipole. The magnetic field density is defined by the equation

$$\mathbf{B} = \frac{\mu_0 M}{4\pi R^3}(2\mathbf{a}_R \cos\varphi + \mathbf{a}_\varphi \sin\varphi), \tag{4.30}$$

where R is the distance from the center of the coil, φ is the polar angle (the coil axis is parallel to $\varphi = 0$), $\mathbf{a}_R$ and $\mathbf{a}_\varphi$ are the radial and tangential unit vectors, and M is the magnetic dipole

$$M = NI\pi b^2, \tag{4.31}$$

Fig. 4.18 Circular electric loop with radius b and current I

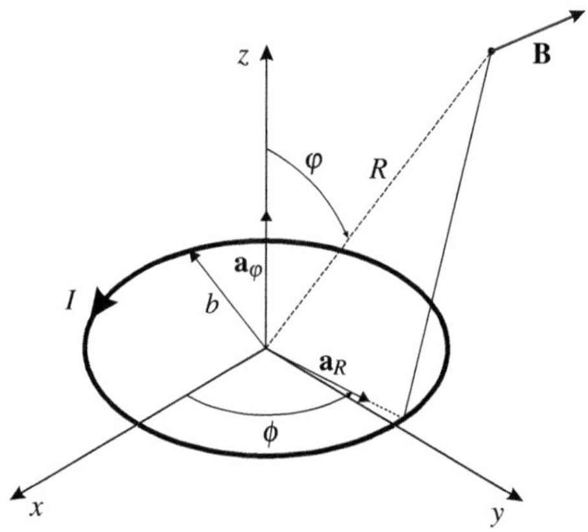

where N is the number of the windings in the coil.

To understand the electromagnetic principle of motion tracking, it is necessary to first understand the phenomenon of coupling. Coupling occurs between a device that emits a signal (called the *source*) and a device that receives the signal (called the *sensor*). The properties of coupling are important for the design of measurement devices that allow objects' poses to be determined. Let's assume that the coil in Fig. 4.18 represents a *source*. The magnetic field density around the coil is approximately determined by Eq. (4.30). We now insert a second coil into this field. The new coil represents the *sensor*. If the *source* generates a changing magnetic field, a voltage is induced in the *sensor* as a function of the magnetic field. The induced voltage changes with the pose of the *sensor* relative to the source. Two examples of the *sensor's* pose relative to the *source* are shown in Fig. 4.19. For the example of coaxial coupling, the *sensor* is in a pose determined with $\varphi = 0°$ with regard to Fig. 4.18. Inputting this value into Eq. (4.30) gives a factor of 2 inside the parentheses, thus defining coaxial coupling as $c = 2$. For the example of coplanar coupling,

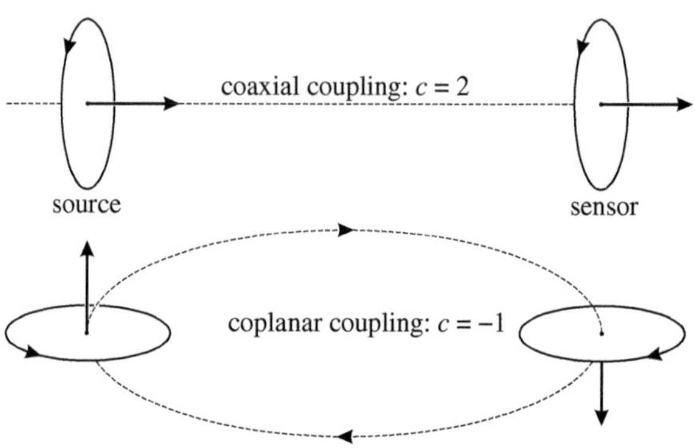

Fig. 4.19 The coupling of an electromagnetic field

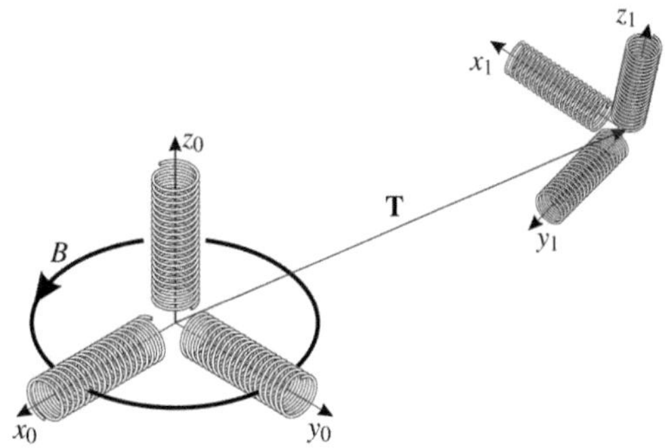

Fig. 4.20 Source of the electromagnetic field and sensor used to determine pose **T**

the *sensor* is in a pose determined with $\varphi = 90°$ with regard to Fig. 4.18. Inputting this value into Eq. (4.30) gives a factor of 1 inside the parentheses. However, since the direction of the magnetic field inside the *sensor* is the opposite of the direction of the magnetic field of the *source*, coaxial coupling is defined with $c = -1$. The coupling factor changes nonlinearly in all intermediate poses, allowing the geometric relations between the *source* and *sensor* to be determined.

Naturally, using a single coil as the *source* and a single coil as the *sensor* does not allow us to calculate the pose of the *sensor* in space. For that, we need three orthogonal coils on the side of both the source and the receiver (Fig. 4.20). Using three orthogonal coils, the source sequentially generates a changing magnetic field in each coil. These fields generate currents in the sensor's coils via coupling. Measuring currents in the sensor allows us to determine the relative pose **T** of the source and sensor. The sensor's signal depends both on the distance between the source and sensor and on their relative orientation. Thus, all six degrees of freedom of the sensor's pose can be determined.

All six degrees of freedom of the *sensor* relative to the *source* are shown in Fig. 4.21. Position is defined by spherical coordinates (R, α, β) while orientation is defined by coordinates (ψ, ϕ, θ). The triaxial electromagnetic dipole represents the reference coordinate system. The *source* generates a temporally multiplexed sequence (coils of the *source* are triggered one after the other) of electromagnetic fields that are sensed by the triaxial magnetic *sensor*. The algorithm used to calculate the pose of the *sensor* is beyond the scope of this work, but can be found in [6].

Finally, a few strengths and weaknesses of tracking based on the electromagnetic principle should be mentioned. Sensors based on this principle are compact, light and relatively cheap. The working area of the system is limited by the decreasing field strength as a function of distance from the *source*. The working area also may not contain any ferromagnetic materials that could distort the magnetic field. The system's advantage is that, unlike optical tracking, it does not require a line of sight between the *source* and *sensor*. The sampling frequency is limited by the temporally multiplexed generation of magnetic fields. The magnetic field in a *source's* coil must dissipate completely before a magnetic field can be generated in the next coil. It

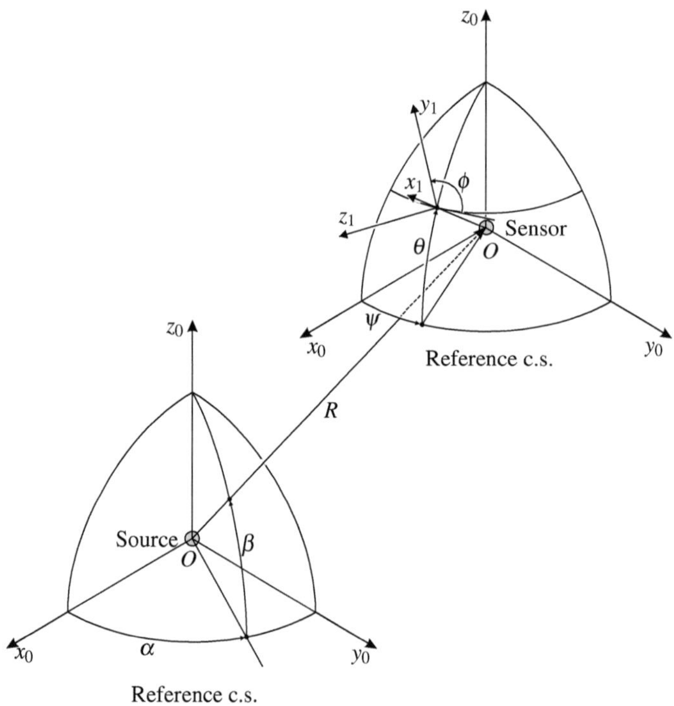

Fig. 4.21 Measuring an object's pose based on the electromagnetic principle

is also necessary to find a compromise between the working area, resolution and accuracy of the system as well as the sampling frequency.

4.1.8 Inertial Principle

The inertial principle is based on inertial measurement systems that combine gyroscopes (angular velocity sensors), accelerometers and magnetometers (which usually measure orientation relative to the Earth's magnetic field). The inertial measurement system works similarly to the inner ear that senses the head's orientation. In principle, inertial systems can measure all six degrees of freedom of an object's pose, though the nonideal sensor outputs represent significant technical challenges. The inertial measurement principle was first used on ships, planes and satellites long before the idea of virtual reality was realized. The basic concept of inertial motion tracking is shown in Fig. 4.22 [3]. The gyroscope is used to calculate the object's orientation while the accelerometer can be used to calculate the object's position using double integration (taking gravitational acceleration into account).

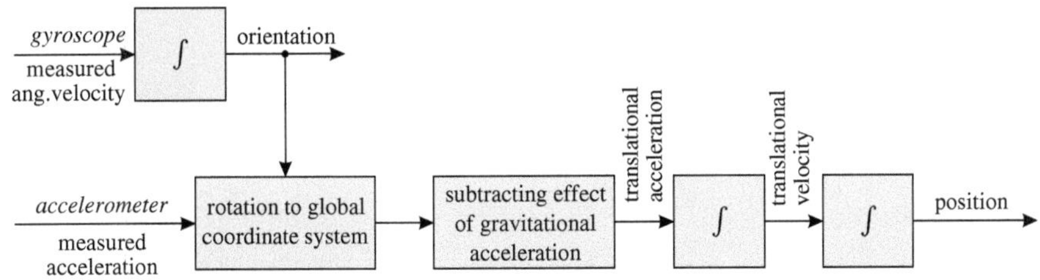

Fig. 4.22 The concept of inertial motion tracking

The first inertial systems were not suitable for use in virtual reality, as the gyroscopes and accelerometers were large mechanical devices that could not be attached to a human. The inertial measurement principle became interesting for virtual reality only with the development of microelectromechanical devices. Microelectromechanical (MEMS) inertial sensors, which include accelerometers and gyroscopes, are one of the most important silicon-based sensors. The greatest demand for them comes mainly from the automotive industry, where they are used to activate safety systems (e.g. air bags), to control vehicle stability and to electronically control the suspension. However, inertial sensors are also used in a variety of other applications where small and cheap sensors are needed. They are used in biomedical applications for human activity recognition, in cameras for image stabilization, in mobile devices and sports goods. The industry uses them for robotics and vibration control while the military uses them to guide missiles. Accelerometers with a high sensitivity are important for autonomous control and navigation, for seismometers and for satellite stabilization in space.

MEMS gyroscopes allow an object's angular velocity to be measured. They are frequently used together with accelerometers and are thus found in most of the same applications.

MEMS magnetometers are not inertial measurement systems on their own, but are frequently used together with accelerometers and gyroscopes and thus included in this group. They are usually used to measure orientation with respect to the Earth's magnetic field.

4.1.8.1 Accelerometer

A basic accelerometer consists of a mass m attached to a rigid housing using springs with the spring coefficient k. Additionally, the damping d affects the movement of the mass. An accelerometer can thus be modeled as a second-order mass-damper-spring system. Such a model is shown in Fig. 4.23. External acceleration causes the housing to move a distance of x relative to the mass, which causes a change in mechanical tension in the elastic springs that support the mass. Both the movement of the mass relative to the housing and the mechanical tension in the springs can be used to measure acceleration. In static conditions, an accelerometer measures

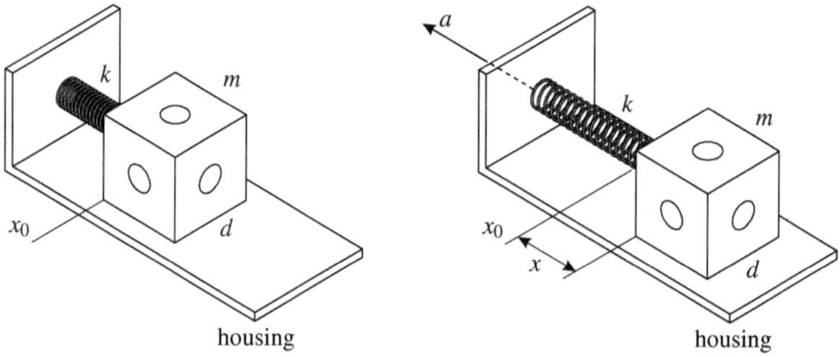

Fig. 4.23 The principle of an accelerometer: static conditions (*left*) and during translational acceleration (*right*)

gravitational acceleration. In dynamic conditions, it measures a vector sum of gravitational acceleration and acceleration due to movement. As we shall see, these sensor characteristics can be used to measure not only translational movement, but also tilt or orientation.

Using Newton' second law and the mechanical model of the accelerometer, it is possible to describe the system dynamics

$$m\ddot{x} + d\dot{x} + kx = ma. \tag{4.32}$$

The product ma in the above equation can be treated as a force acting on the mass. Due to the force, the mass moves as described by the dynamics of the system and its parameters m, d and k. Equation (4.32) can be written in the form of a transfer function

$$H(s) = \frac{x(s)}{a(s)} = \frac{1}{s^2 + \frac{d}{m}s + \frac{k}{m}} = \frac{1}{s^2 + \frac{\omega_r}{Q}s + \omega_r^2}, \tag{4.33}$$

where $\omega_r = \sqrt{k/m}$ is the system's resonance frequency and $Q = \sqrt{km}/d$ is the quality factor of the system. The resonance frequency can be increased by increasing the spring constant k or decreasing the mass m while the quality factor can be increased by decreasing damping d or increasing the mass m and spring constant k.

In general, a single accelerometer housing can contain elements to measure acceleration along all three translational axis. An implementation of such a system that uses the capacitive principle to measure displacement is shown in Fig. 4.24. The typical characteristic parameters of accelerometers include sensitivity, operating range, frequency response and resolution.

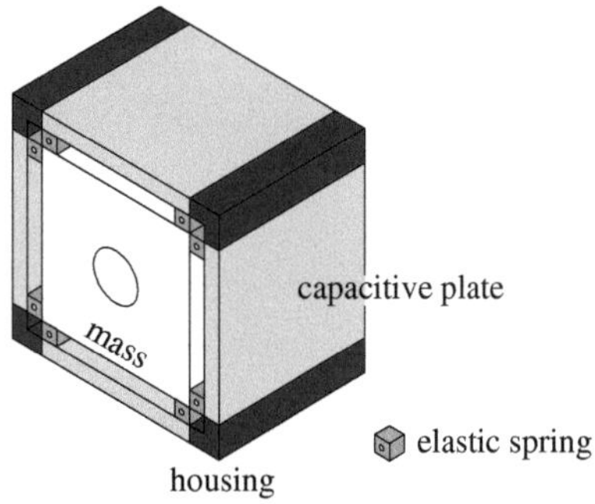

Fig. 4.24 Multiaxial accelerometer with mass and capacitive plates

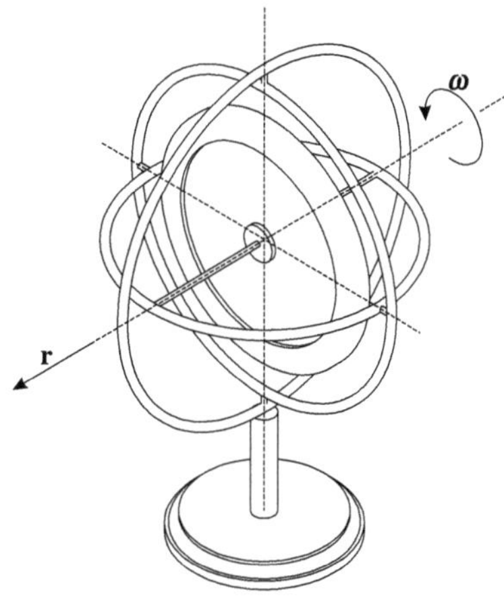

Fig. 4.25 Classic gyroscope with a spinning wheel: the vector **r** defines the axis of rotation while the angular momentum **L** is defined as the product of the moment of inertia $\boldsymbol{I}$ and the angular velocity of the spinning wheel $\boldsymbol{\omega}$

4.1.8.2 Gyroscope

Gyroscopes are devices that measure angular velocities. They are generally divided into three major categories: (1) mechanical gyroscopes, (2) optical gyroscopes and (3) gyroscopes based on a vibrating mass. Mechanical gyroscopes are based on the principle of conservation of angular momentum (Fig. 4.25). Newton's second law states that angular momentum is conserved while no external torques act upon the system. The basic equation that describes a gyroscope is

$$\boldsymbol{\tau} = \frac{d\mathbf{L}}{dt} = \frac{d(\boldsymbol{I}\boldsymbol{\omega})}{dt} = \boldsymbol{I}\boldsymbol{\alpha}, \tag{4.34}$$

where $\boldsymbol{\tau}$ is the torque acting on the gyroscope, $\mathbf{L}$ is the gyroscope's angular momentum, $\boldsymbol{I}$ is the gyroscope's moment of inertia, and $\boldsymbol{\omega}$ and $\boldsymbol{\alpha}$ are the angular velocity and angular acceleration.

Optical gyroscopes are based on a laser light source and interferometry. Their main advantage is that they have no moving parts, so they are not vulnerable to mechanical wear and signal drift.

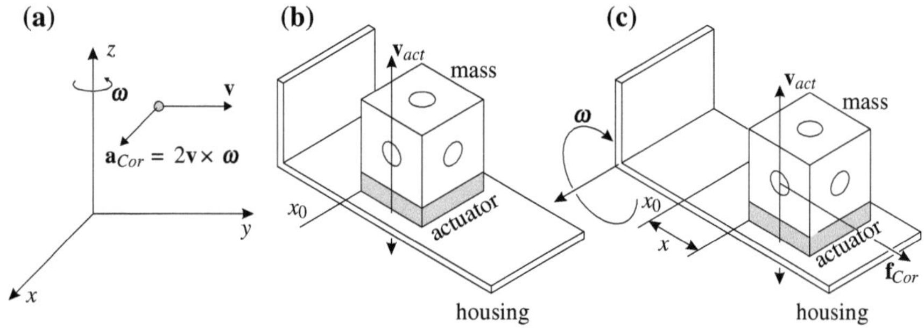

Fig. 4.26 Operating principle of a gyroscope: **a** Coriolis acceleration, **b** gyroscope in static conditions and **c** gyroscope during rotation

Mechanical and optical gyroscopes are not suitable for attachment to humans. However, gyroscopes based on a vibrating mass are small and cheap enough to use in motion tracking. The principle is based on a vibrating mass that is affected by Coriolis acceleration due to rotation. This acceleration causes secondary vibrations perpendicular to the direction of primary vibrations and the angular velocity vector. By measuring the secondary vibrations, it is possible to determine the object's angular velocity. Figure 4.26a shows the basic principles of such a gyroscope. A particle moving with velocity $\mathbf{v}$ is affected by acceleration $\mathbf{a}_{Cor}$ due to the rotation of the system at angular velocity $\boldsymbol{\omega}$.

In a gyroscope based on a vibrating mass, the actuator causes vibrations that move the mass with velocity $\mathbf{v}_{act}$. There is no additional movement of the mass as long as angular velocity is equal to zero (Fig. 4.26b). However, the angular velocity $\boldsymbol{\omega}$ (Fig. 4.26c) causes a Coriolis force

$$\mathbf{f}_{Cor} = 2m(\mathbf{v}_{act} \times \boldsymbol{\omega}), \quad (4.35)$$

that creates additional movement of the mass in the direction of the force. By measuring the displacement x, it is possible to calculate the angular velocity of the sensor housing.

In general, a single housing can contain components that measure angular velocities around all three axes of rotation. Typical characteristic parameters of gyroscopes include sensitivity, operating range and resolution.

4.1.8.3 Magnetometer

A magnetometer measures the strength of an external magnetic field, be it natural or artificial. It can thus be used as an orientation sensor, usually in combination with accelerometers and gyroscopes. Since a sensor for use in virtual reality must be small and cheap, let's more closely examine a method of measuring magnetic field strength based on the magnetoresistive effect: the property of some conductive

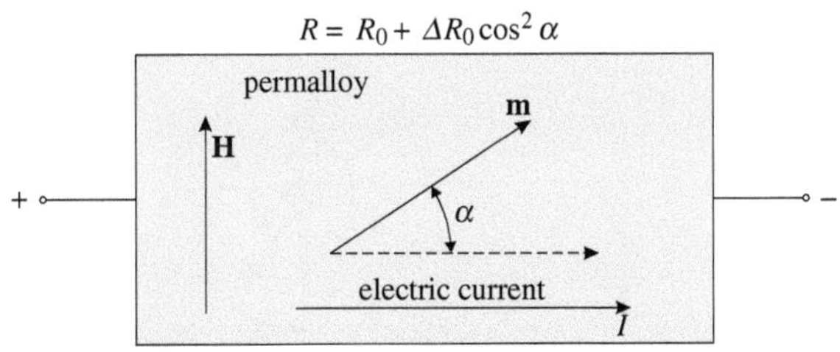

Fig. 4.27 Magnetoresistive principle of a magnetometer

materials that their resistance changes due to an external magnetic field. The basic operating principle of the sensor is shown in Fig. 4.27.

The figure shows a slice of the ferromagnetic material permalloy. Assuming that there is no external magnetic field, the magnetization vector of permalloy $\mathbf{m}$ is parallel to the direction of the current (in our case, left to right). If the sensor is placed into a magnetic field with strength $\mathbf{H}$ such that the field is parallel to the plane of the permalloy and perpendicular to the direction of the current, the internal magnetization vector of the permalloy $\mathbf{m}$ turns by an angle α. Consequently, the resistance of the material R changes as a function of the angle α

$$R = R_0 + \Delta R_0 \cos^2\alpha, \tag{4.36}$$

where R_0 and ΔR_0 are properties of the material that give optimal sensor characteristics. Measuring the resistance R allows measurement of the sensor's angle relative to the external magnetic field, which can be used to determine the object's spatial orientation.

4.1.8.4 Inertial Measurement Unit

Individual sensors (accelerometer, gyroscope or magnetometer) can be used as a basis for user motion tracking. We will analyze the use of a two-axis accelerometer (measures accelerations along two perpendicular axes) on the example of a rigid pendulum. Both sensors give the measured quantities in their own coordinate systems, which are attached to the center of the sensor and have their axes parallel to the axes (x, y) of the coordinate system attached to the pendulum. Figure 4.28a shows a stationary pendulum while Fig. 4.28b shows a swinging pendulum. We are interested in the pose of the pendulum relative to the reference coordinate system (x_0, y_0, z_0). Since the pendulum is only swinging around axis z, we are actually only interested in angle φ.

We first analyze stationary conditions. Since the angular velocity of a stationary pendulum is equal to zero, the gyroscope's output is also zero and the gyroscope tells us nothing about the pendulum's orientation. However, we can see that the accelerometer still measures the gravitational acceleration. Since the accelerometer

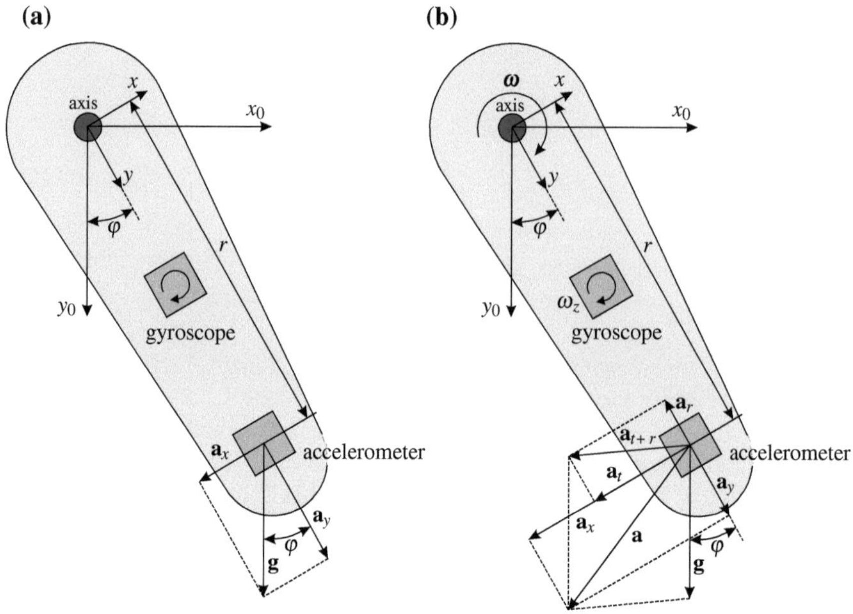

Fig. 4.28 Example of using an inertial measurement system to measure the angle of a pendulum. **a** Stationary pendulum. **b** Swinging pendulum

is at an angle of φ relative to the gravitational field, two acceleration components are measured: $\mathbf{a}_x$ and $\mathbf{a}_y$. The vector sum of both components gives the gravitational acceleration. Figure 4.28a shows that the angle between vectors $\mathbf{g}$ and $\mathbf{a}_y$ is equal to φ. Since the scalar values of accelerations g and a_y are known, we can now determine the pendulum angle

$$\varphi = \arccos \frac{a_y}{g}. \tag{4.37}$$

The accelerometer thus allows the pendulum's angle to be measured in stationary conditions. For this reason, accelerometers are frequently used as inclinometers.

The conditions in a swinging pendulum are quite different. Since swinging is an accelerated rotational movement, the accelerometer is affected not only by the gravitational acceleration $\mathbf{g}$, but also by centripetal acceleration

$$\mathbf{a}_r = \boldsymbol{\omega} \times (\boldsymbol{\omega} \times \mathbf{r}) \tag{4.38}$$

and tangential acceleration

$$\mathbf{a}_t = \dot{\boldsymbol{\omega}} \times \mathbf{r}. \tag{4.39}$$

The total acceleration acting on the accelerometer is thus

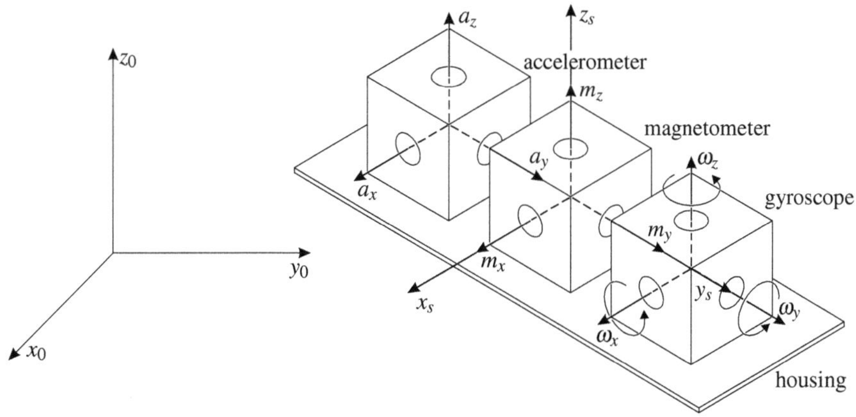

Fig. 4.29 Inertial measurement unit consisting of three accelerometers, three gyroscopes and three magnetometers

$$\mathbf{a} = \mathbf{g} + \mathbf{a}_r + \mathbf{a}_t. \tag{4.40}$$

The equation used to calculate angle in stationary conditions (4.37) is no longer valid, so the accelerometer cannot be used to calculate the angle of a swinging pendulum. However, the output of the gyroscope, which measures the angular velocity of the pendulum, is now also available. Since the angle of the pendulum can be calculated as the temporal integral of angular velocity, the following relation can be stated

$$\varphi = \varphi_0 + \int \omega_z dt, \tag{4.41}$$

where the initial orientation of the pendulum φ_0 must be known.

The given example makes it clear that an accelerometer is suitable for orientation measurements in static or quasistatic conditions while a gyroscope is suitable for orientation measurements in dynamic conditions. However, two weaknesses of accelerometers and gyroscopes must be mentioned. An accelerometer cannot be used to measure angles in a horizontal plane, as the output of the sensor is zero when its axis is perpendicular to the direction of gravity. Furthermore, neither the gyroscope's nor the accelerometer's output is ideal. In addition to the measured quantity, the output includes an offset and noise. Integrating the offset causes a linear drift, so Eq. (4.41) does not give an accurate pendulum orientation measurement.

Due to the weaknesses of the individual sensors, it is common to combine three perpendicular accelerometers, three perpendicular gyroscopes and three perpendicular magnetometers into a single system referred to as an inertial measurement unit (IMU, Fig. 4.29). An IMU allows measurement of the orientation of the sensor's coordinate system (x_s, y_s, z_s) relative to the reference coordinate system (x_0, y_0, z_0) in three degrees of freedom. For short intervals of time (where drift is not danger-

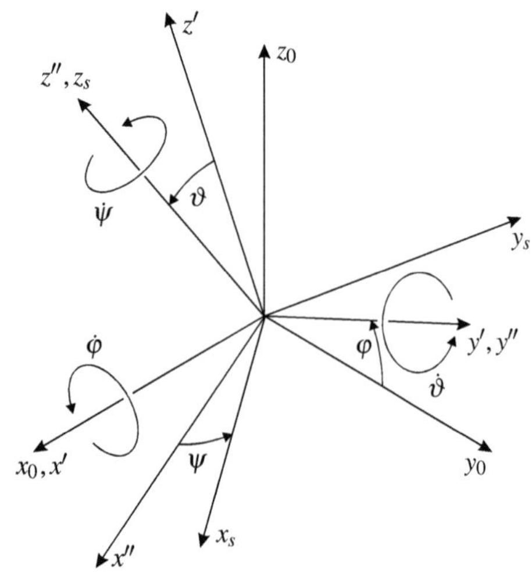

Fig. 4.30 Angular velocities and orientation given in *XYZ* Euler angles

ous), integration of acceleration also allows calculations of translational velocities and position (x_s, y_s, z_s) relative to (x_0, y_0, z_0).

A gyroscope measures the angular velocity vector $\dot{\boldsymbol{\phi}} = \left[\dot{\varphi}\ \dot{\vartheta}\ \dot{\psi}\right]^T$, which is defined in the gyroscope's local coordinate system (Fig. 4.30). Integrating the angular velocity vector $\dot{\boldsymbol{\phi}}$ over time gives the sensor orientation

$$\boldsymbol{\phi} = \begin{bmatrix} \varphi \\ \vartheta \\ \psi \end{bmatrix} = \int_0^t \dot{\boldsymbol{\phi}} d\tau \tag{4.42}$$

written in the form of *XYZ*-Euler angles as seen in Fig. 4.30. Vector $\boldsymbol{\phi}$ gives the orientation of the coordinate system (x_s, y_s, z_s) with regard to coordinate system (x_0, y_0, z_0).

In static and quasistatic conditions, the accelerometer allows measurement of rotations around the x_0 and y_0 axes of the reference coordinate system. Since the gravity vector is parallel to axis z_0, the accelerometer cannot sense rotation around this axis. For this purpose, we can use a magnetometer, which also allows measurement of rotation around the z_0 axis (think of how a compass works). Combining an accelerometer and magnetometer thus gives an estimate of sensor spatial orientation, but such measurements are suitable only for quasistatic conditions. In dynamic conditions, an additional acceleration acts on the accelerometer and prevents it from being used as a tilt sensor.

Combining the best properties of an accelerometer, gyroscope and magnetometer can give us an accurate and reliable measurement of spatial orientation. This is done through sensor integration, which can be done e.g. using the Kalman filter [7]. Sensor integration is beyond the scope of this work, so let's look only at the basic concept as illustrated in Fig. 4.31.

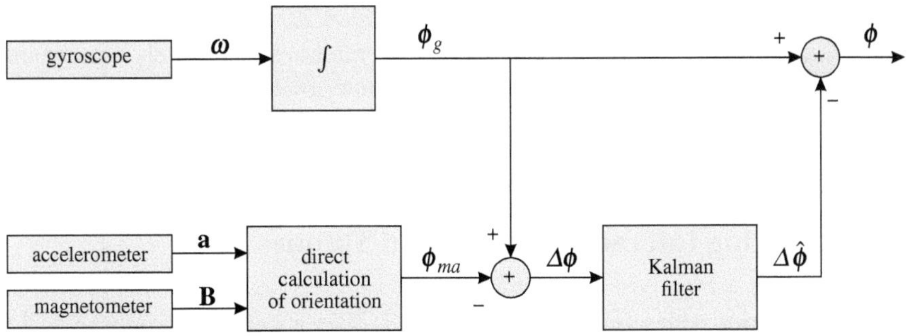

Fig. 4.31 The concept of sensor integration using a Kalman filter

The angular velocity measured by the gyroscope is first integrated, giving an initial estimate of orientation $\boldsymbol{\phi}_g$. Measurements from the accelerometer and magnetometer are used to directly calculate the sensor orientation with regard to the gravity and magnetic field vectors $\boldsymbol{\phi}_{ma}$. Both orientation estimates are compared, and the difference between the two $\Delta\boldsymbol{\phi}$ is input to a Kalman filter that uses a model of the process to calculate the correction $\Delta\hat{\boldsymbol{\phi}}$. The model of the process takes into account the properties of measurements with accelerometers (suitable for quasistatic conditions) and gyroscopes (suitable for dynamic conditions). The calculated correction is used to adjust the estimated orientation obtained from the gyroscopes $\boldsymbol{\phi}_g$.

The transfer function of the system in Fig. 4.31 can be written as follows

$$\boldsymbol{\phi} = \frac{s^2}{s^2 + k_1 s + k_2} \frac{\boldsymbol{\omega}(s)}{s} + \frac{k_1 s + k_2}{s^2 + k_1 s + k_2} \boldsymbol{\phi}_{ma}(s) \tag{4.43}$$

or in a simpler form

$$\boldsymbol{\phi} = \mathbf{G}(s) \frac{\boldsymbol{\omega}(s)}{s} + (1 - \mathbf{G}(s))\, \boldsymbol{\phi}_{ma}(s). \tag{4.44}$$

The latter equation clearly shows that the Kalman filter weighs two different sources of information (integrated gyroscope signal and absolute measured orientation) that complement each other. The function $\mathbf{G}(s)$, which filters the integrated gyroscope signal, acts as a high-pass filter. In other words, at high frequencies or during rapid motions the output is equal to the integrated gyroscope signal. The function $1-\mathbf{G}(s)$, on the other hand, filters the absolute measured orientation and acts as a low-pass filter. During slow motions, the filter output thus gives more weight to the absolute angle measurement. Similar findings were seen during the analysis of the pendulum angle.

Taking the calculated orientation into account, we can now subtract the gravitational acceleration from the measured acceleration. The remnant represents the translational acceleration. Integrating this acceleration allows us to estimate the velocity and position of an object in space as shown in Fig. 4.22. Of course, we must take into account that the remaining offset value (we can never completely remove the effect of gravity) causes the velocity and position to drift. This method can thus only be used over brief time periods.

4.2 Measuring Interaction Forces and Torques

During interaction with a virtual environment, it is often not enough to only know the pose and movement of the user. We are also interested in the interaction forces. An example of such a situation is the measurement of the ground reaction force during balancing tasks. This section thus presents two methods of measuring forces and torques between the user and virtual environment. We wish to know the point on which the total force and torque act as well as their amplitude and direction.

4.2.1 Relationship Between Interaction Force and Point of Contact

Contact between a human and an object in the environment can be described with the position of the point of contact **p** and the interaction force **f**. Six variables are thus required: x_p, y_p, z_p, f_x, f_y in f_z. We must track the point of contact in the environment's coordinate system as well as measure all three components of the interaction force. These six values can be determined in different ways.

Let's first see how we can determine the position of the point of contact if we know the interaction force **f** and its torques acting around the axes of the basic coordinate system. The geometric conditions are shown in Fig. 4.32. The force **f** can be divided

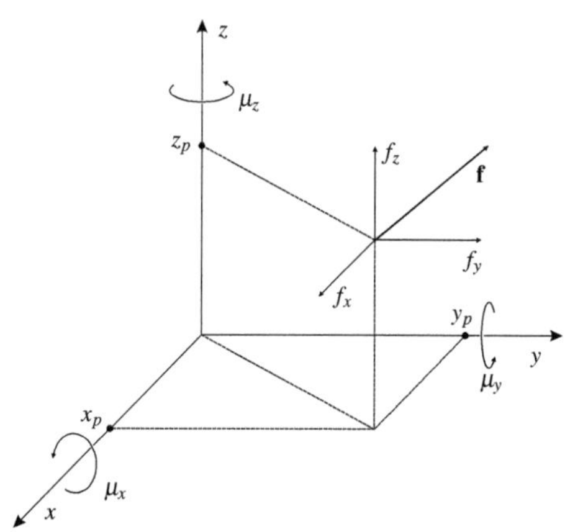

Fig. 4.32 Relationship between forces and torques

into three components: f_x, f_y and f_z. These forces create the following torques with regard to the axes of the basic coordinate system

$$
\begin{aligned}
\mu_x &= f_z y_p - f_y z_p \\
\mu_y &= f_x z_p - f_z x_p \\
\mu_z &= f_y x_p - f_x y_p.
\end{aligned} \tag{4.45}
$$

If we multiply the individual equations in (4.45) by f_x, f_y in f_z and then sum them together, we get

$$f_x \mu_x + f_y \mu_y + f_z \mu_z = 0. \tag{4.46}$$

This means that equations (4.45) are redundant. Only two coordinates of the point of contact can be calculated from them. However, this is usually enough since one coordinate of the point of contact is generally given.

The interaction force between the user and the environment is generally measured either with force sensors or force plates.

4.2.2 Force and Torque Sensor

A force and torque sensor is usually placed between the user and object in the area where we wish to measure interaction with the user.

A force and torque transducer generally has a cross-shaped mechanical structure as seen in Fig. 4.33. The ends of the cross are attached to an object in the environment while the grip comes from an opening in the center. Eight pairs of semiconductive measurement strips are affixed to the ends of the cross. These strips allow the measurement of deformations from w_1 to w_8. Each opposing pair of measurement strips is connected to a measurement bridge. When force is applied to the sensor, it is possible to record eight analog voltages proportional to the forces marked in the figure.

A calibration procedure allows us to determine the elements of a 6×8 calibration matrix that converts the measured analog values to force and torque vectors

$$\left[f_x \; f_y \; f_z \; \mu_x \; \mu_y \; \mu_z\right]^T = \mathbf{K}\left[w_1 \; w_2 \; w_3 \; w_4 \; w_5 \; w_6 \; w_7 \; w_8\right]^T, \tag{4.47}$$

where

$$\mathbf{K} = \begin{bmatrix} 0 & 0 & K_{13} & 0 & 0 & 0 & K_{17} & 0 \\ K_{21} & 0 & 0 & 0 & K_{25} & 0 & 0 & 0 \\ 0 & K_{32} & 0 & K_{34} & 0 & K_{36} & 0 & K_{38} \\ 0 & 0 & 0 & K_{44} & 0 & 0 & 0 & K_{48} \\ 0 & K_{52} & 0 & 0 & 0 & K_{56} & 0 & 0 \\ K_{61} & 0 & K_{63} & 0 & K_{65} & 0 & K_{67} & 0 \end{bmatrix} \tag{4.48}$$

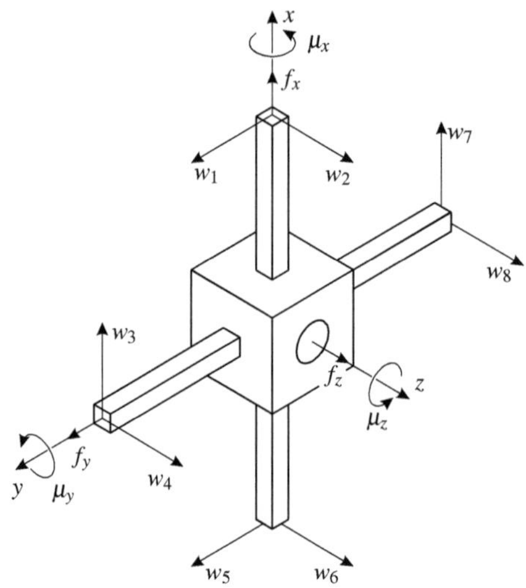

Fig. 4.33 Force sensor concept

is the calibration method whose elements K_{ij} are constant and represent the gains of the individual measurement bridges.

4.2.3 Force Plate

The interaction force between the user and environment can also be measured with a force plate. Such a plate contains force transducers in all four corners to measure the horizontal and vertical components of the measured reaction force (Fig. 4.34). The coordinate system's point of origin is placed in the center of the plate. Horizontal forces in the plate corners are marked as α_i while vertical forces are marked as β_i.

Interaction forces and torques are obtained from the forces measured in the corners of the force plate

$$\begin{aligned} f_x &= \alpha_3 - \alpha_1 \\ f_y &= \alpha_4 - \alpha_2 \\ f_z &= \beta_2 + \beta_4 - \beta_1 - \beta_3 \\ \mu_x &= (-(\beta_1 + \beta_4) + (\beta_2 + \beta_3))\,\tfrac{l}{2} \\ \mu_y &= (-(\beta_3 + \beta_4) + (\beta_1 + \beta_2))\,\tfrac{l}{2} \\ \mu_z &= (\alpha_1 + \alpha_2 + \alpha_3 + \alpha_4)\,\tfrac{l}{2}. \end{aligned} \tag{4.49}$$

The plane of the force plate has the vertical coordinate $z_p = 0$. It is then possible to calculate the coordinates of the point $\mathbf{p}$ onto which the reaction force $\mathbf{f}$ acts

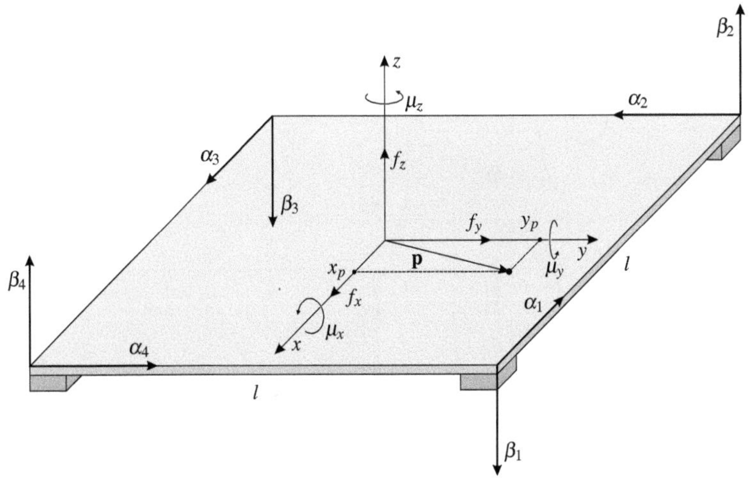

Fig. 4.34 Force plate for measurement of forces and torques

$$\begin{aligned} x_p &= -\mu_y/f_z \\ y_p &= -\mu_x/f_z. \end{aligned} \tag{4.50}$$

Let's say that the goal is to determine the position of the vertical projection $\overline{\mathbf{COG}}$ of the center of gravity **COG** of a person standing on the force plate (Fig. 4.35). The sensors give information about the quantities f_x, f_y, f_z, μ_x, μ_y in μ_z, thus allowing the projection of the center of gravity to be calculated from the Eq. (4.50).

4.3 Motion Tracking

Motion tracking allows the user's pose and activities in a virtual reality system to be measured. The body parts that need to be measured depend on each application's specific requirements. A *gesture* is defined as a specific motion that occurs at a certain time. Gestures can be used for intuitive interaction with the virtual environment. Usually, body motion tracking focuses on the head, trunk, hands and fingers, or feet.

4.3.1 Head

Especially *head-mounted displays* (usually visual or audio displays attached to the head) require head orientation to be measured, since the image on the display depends on the relative orientation of the head. *Head pose information* is necessary to create precise views of the virtual environment. In the case of stationary displays, it is necessary to know the eye position relative to the display while gaze direction is irrelevant. Six head pose degrees of freedom must be measured for appropriate

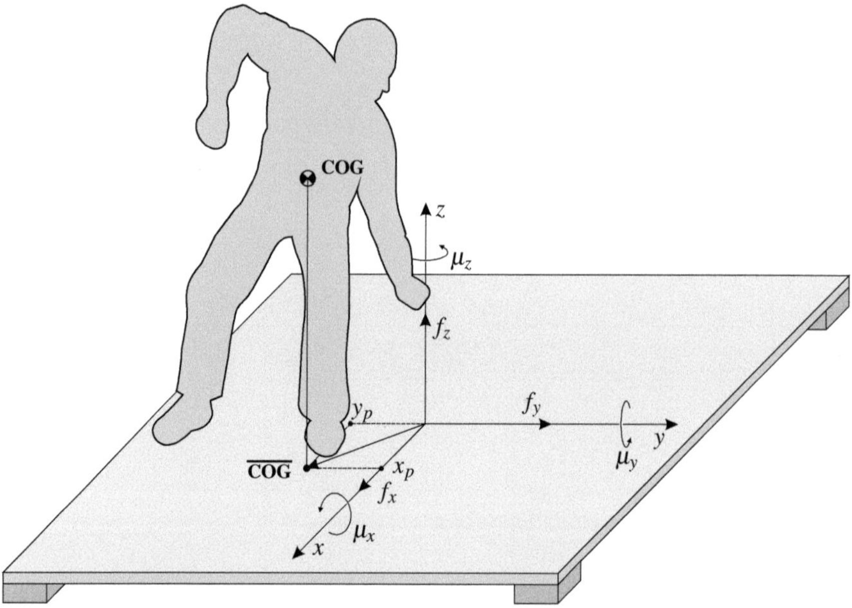

Fig. 4.35 Determining the vertical projection of the center of gravity of a person standing on a force plate

stereoscopic display. In the case of head-mounted displays, information about head orientation is needed for appropriate display (e.g., when turning our head to the left, we expect to see the things located on the left).

Pose measurements (especially head orientation) are most often done with inertial measurement systems (Fig. 4.36). For head-mounted displays, the measurement system is usually already integrated into the display. Of course, it is also possible to use ultrasonic, optical or electromagnetic methods to measure pose and head movement.

4.3.2 Hand and Finger Tracking

Hand and finger tracking allows a user to interact with the virtual world. In multi-user environments, gesture recognition can be used to communicate between subjects. Mechanical, optical, ultrasonic, electromagnetic or inertial sensors can be used for arm tracking. Finger tracking, however, is limited by the maximum size of the sensor and by the number of degrees of freedom that should be measured. Special gloves are thus usually used to measure finger movement. These gloves are equipped with sensors (goniometers, optic fibers) that are placed along the fingers and measure finger joint angles (Fig. 4.37). A goniometer is a sensor whose resistance changes as it is bent, thus allowing joint angles to be measured via resistance changes. Bent

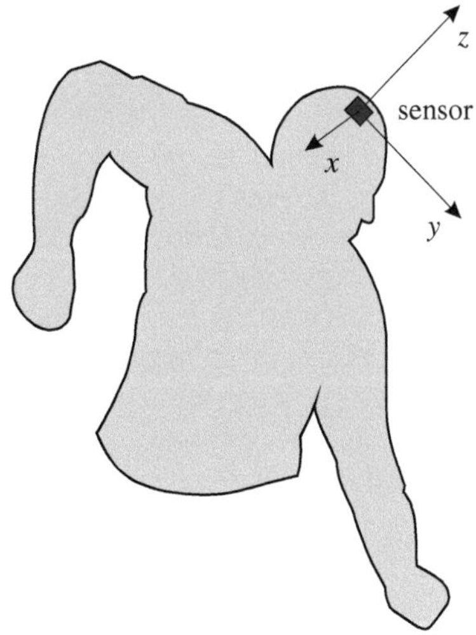

Fig. 4.36 Head motion tracking using an inertial measurement system

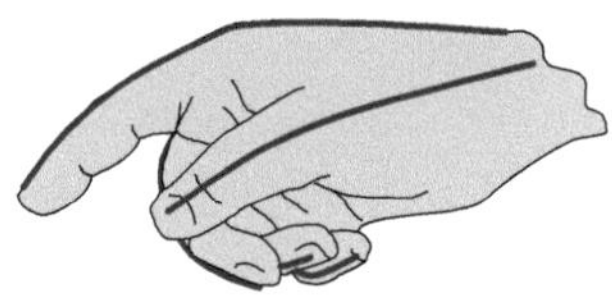

Fig. 4.37 Finger tracking using resistance (goniometers) or optical (optic fibers) sensors. Sensors are attached along the finger and measure joint angles

optic fibers, on the other hand, change the amount of light that can pass through them, thus allowing joint angles to be calculated from the amount of light.

4.3.3 Eyes

Studies have shown that the eyes are the human's most frequently moved external part of the body. The constant motion occurs due to the huge amount of visual data coming from the environment. The eye does not study the entire available visual scene in detail, but focuses on smaller bits of information that are examined rapidly one after another. By integrating individual parts of the visible world, the human creates an interconnected whole [8]. Small eye movements and thus changes in perceived light occur even when the eye is focused on a small detail in the environment. Since the nerve endings in the eye can detect only changes in light, this is the only way we can see stationary objects.

Eye movement consists of rapid movements called saccades and stationary periods called fixations. Saccades occur 3–4 times a second, with each one lasting 20–200 ms. During rapid movements, the eye perceives barely any visual information. Between individual saccades, the eye briefly stops and focuses on a small part of the visual scene [8]. Eye movement characteristics such as the number of fixations, the duration of an individual fixation, the number of saccades and the saccade amplitudes contain important information about visual attention.

The concept of visual attention has been intensively studied for over a hundred years. Its beginnings can be found in psychology, where scientists wished to determine why humans focus attention on only one object among many and why some objects are viewed longer than others. Eye tracking methods gradually became well-developed and spread from psychology to other fields. They can be found in branches of medicine such as neurology and ophthalmology, in marketing and advertising, for user interface evaluation and for eye tracking studies in everyday tasks such as reading, driving or shopping. Eye positions can also serve as an input to a system, creating the principle of gaze-based interaction.

Eye tracking systems measure and perceive four different eye-related phenomena: (1) eye rotation within the eye sockets, (2) gaze direction (taking head and eye movements into account), (3) blinking and (4) pupil size. Important properties of eye trackers include sampling frequency, eye position measurement accuracy, and robustness with regard to changing light conditions and environmental noise.

The first important issue is how quickly and how accurately eye position can be determined. Eye trackers for medical and psychological research sample at 500 Hz and can achieve accuracies of up to 1' visual angle. For user interface research, sampling frequencies of 60 Hz and accuracies of 1° are generally sufficient [9]. On the other hand, the eye tracker must be as unobtrusive as possible, allowing the subject to move freely around the room if possible.

Depending on the measured parameter, eye tracking systems can be divided into those that measure eye position relative to the head and those that measure the gaze direction (also called point of regard). The latter are used primarily in applications where it is necessary to recognize visual elements that attract a user's attention [10].

In general, modern computer-aided eye trackers can be divided into three categories with regard to the hardware and measurement principle: (1) electrooculography, (2) search coils and (3) videooculography. We will focus only on videooculography, which is based on capturing a visual image of the face or the eye region. Image processing algorithms allow the gaze direction to be determined based on the eye's optical properties.

Simple algorithms search for only one reference point, most commonly the border between the pupil and the iris or the border between the iris and the sclera (called the limbus). Appropriate image processing algorithms (edge detection) make it easier to find the border between the colored iris and the white limbus. Finding the border between the retina and iris is harder due to the lower contrast, especially in people with dark eyes. For this purpose, we can either additionally illuminate the eye or improve the contrast by equipping the camera with a filter that allows only near-infrared or infrared light to pass. A camera with such a filter can capture an image with a much more pronounced retina. The contrast between the retina and any color of iris thus increases.

If an eye tracker measures two properties of the moving eyes, it is possible to calculate the subject's gaze direction using appropriate calibration. Two appropriate eye properties are the center of the illuminated retina and the corneal reflection

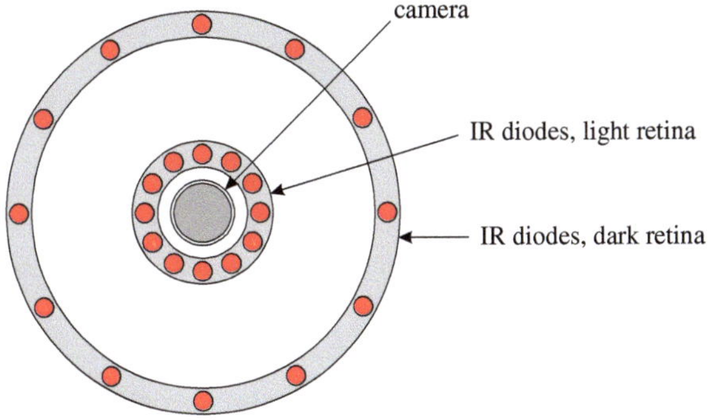

Fig. 4.38 Placement of the camera as well as the *inner* and *outer rings* of infrared light emitting diodes

(reflection of light from the outer surface of the cornea). The reflection occurs due to the additional light source pointed at the subject's eyes. Most typically, infrared light sources (e.g. light emitting diodes) are used to emphasize the retina and cause the reflection which is then used to calculate the eye's current position.

The infrared illumination makes the retina look darker or lighter in the camera's image. It looks lighter if the infrared diodes are fixed to the virtual axis connecting the camera and human eyes. If illumination is not fixed to this axis, the retina looks darker. One method of locating the retina makes use of this light and dark effect. The images captured by the camera are divided into even and odd. Image capturing is synchronized with infrared illumination as shown in Fig. 4.38. The eyes are illuminated with infrared diodes on the camera's axis (inner ring) for even images and with diodes not on the camera's axis for odd images (outer ring). The differences in illumination make the retina look lighter on even images and darker on odd ones. The retina is then detected by thresholding the difference between captured even and odd images [11].

Due to the eye's structure, three reflections are visible in addition to the reflection from the outer surface of the cornea (the most visible one): reflection from the inner surface of the cornea, reflection from the outer surface of the lens and reflection from the inner surface of the lens. These four reflections are collectively known as Purkinje images. The position of the light reflection from the outer surface of the cornea (also called the first Purkinje image) relative to the center of the retina changes as the eyes are moved, but remains relatively unchanged in small head movements. The distance between the two optical characteristics allows the orientation of the eye to be determined. Figure 4.39 shows a light and dark retina as well as the reflection from the outer surface of the cornea. Figure 4.40 shows an example of the left eye of a subject gazing at nine calibration points as well as the relative position of the first Purkinje image with regard to the retina. Most modern eye tracking systems track only two points of the eye: the center of the retina and the reflection from the outer surface of the cornea. A special type of systems called dual-Purkinje image eye trackers simultaneously measure reflections from the outer surface of the cornea and from

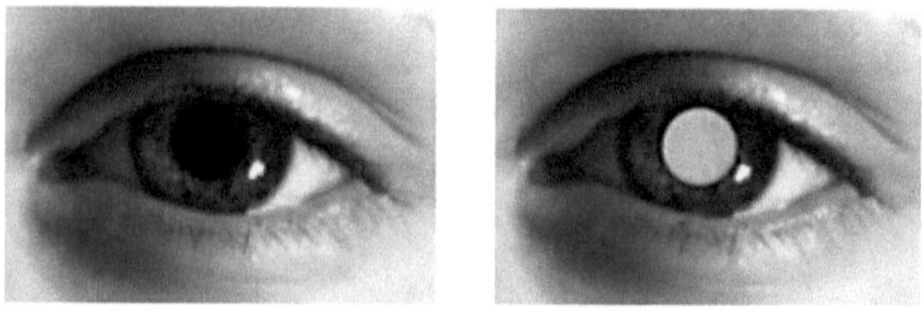

Fig. 4.39 Images of a dark (*left*) and light (*right*) retina as well as the reflection from the outer surface of the cornea (*white spot*)

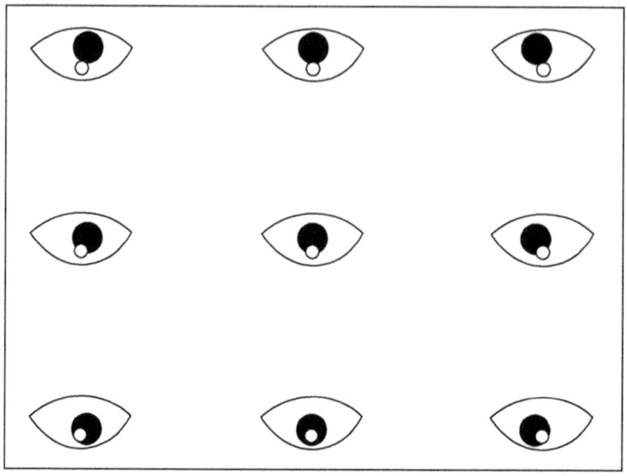

Fig. 4.40 Position of the first Purkinje image (*white spot*) relative to the retina (*black circle*) for a *left eye* gazing at nine calibration points

the inner surface of the lens. They can thus differentiate between translational and rotational eye movement since the two reflections move together during translation, but not during rotation. Such an eye tracking system is very accurate, but requires the subject's head to be fixed [10].

Eye tracking systems also differ with regard to the position of the camera. If we wish to, for instance, measure the region of interest on a computer screen, the camera can be fixed to the table or to the top of the screen. The difficulty comes from finding cameras with adequate magnification and suitable viewing angle. Those with small magnification and wide viewing angle allow the user to perform more head movements, but also capture many objects in the background that can impede eye tracking performance. Cameras with large magnification are much more accurate and usually capture only a single eye, but require the subject to stay still to avoid moving out of the camera's field of view. The best option is a combined system with two cameras and a mechanism that allows the second camera to be moved. The wide-angle camera first approximately locates the eye, and the mechanism turns the large-magnification camera in the appropriate direction. The second camera then takes a high-resolution picture of the eye. The second group of eye trackers are head-mounted systems, which offer a higher accuracy since the camera is always in the same position relative to the eye. However, the camera only tracks eye movements. Such head-mounted systems need to be unobtrusive, lightweight and suitable for a large spectrum of users.

4.3.4 Trunk

Trunk motion tracking can give higher-quality information about body movement direction than information obtained from the head orientation. If the head orientation is used to determine movement direction, it becomes impossible for the subject to look sideways while walking straight ahead.

Since the trunk has a relatively large area, most tracking methods can be used for it. One exception is the mechanical principle, which is more suitable for the extremities.

4.3.5 Legs and Feet

Leg and foot movement is less often tracked, but allows the subject's speed and direction of movement to be measured. Leg motion tracking is usually used when the subject is moving in a large area and thus requires methods that don't impede movement. Examples of such methods are ultrasonic, optical or inertial. Other methods are less suitable for motion tracking in situations with large position changes.

4.4 Physical Input Devices

Physical input devices are usually part of the interface between the user and virtual world. They can be either simple handheld objects or complex platforms. A person operating the physical device gets an impression of the object's physical properties such as weight and texture, thus receiving a type of haptic feedback.

4.4.1 Physical Controls

Physical controls include individual buttons, switches, dials and slides that allow the user to directly affect virtual reality.

4.4.2 Props

Props are physical objects used as interfaces with the virtual world. The prop can be connected to a virtual object and/or have physical controls attached to it. The physical properties of the prop (shape, weight, texture, hardness) often already indicate its use in virtual reality. Such props allow intuitive and flexible interaction with the virtual world. Since it is easy to determine the spatial relationship between two props or between the prop and the user, the user can use this information to better understand the virtual world. The goal of props is to create an interface that allows the user natural manipulation of the virtual world; in other words, to approximate an ideal interface that would barely be noticed by the user.

4.4.3 Platform

As the name suggests, a platform is a large and not very mobile physical structure used as an interface with the virtual world. Just like props, platforms represent a part of the virtual world through real objects that the user can interact with. The platform thus becomes part of the virtual reality system. It can be designed to imitate a real-world device that is simultaneously part of virtual reality, but it can also represent a general space for the user to stand or sit in. One example of a platform is thus the cockpit of an airplane.

4.4.4 Speech Recognition

Speech recognition allows natural communication with a computer system. The use of speech makes the experience of virtual reality more convincing and realistic.

References

1. Rolland JP, Baillot Y, Goon AA (2001) A survey of tracking technology for virtual environments
2. Zhou H, Hu H (2008) Human motion tracking for rehabilitation—A survey. Biomed Signal Process Control 3:1–18
3. Welch G, Foxlin E (2002) Motion tracking: no silver bullet, but a respectable arsenal. IEEE Comput Graph Appl 22:24–38
4. Kuang WT, Morris AS (2000) Ultrasound speed compensation in an ultrasonic robot tracking system. Robotica 18:633–637
5. Shotton J, Fitzgibbon A, Cook M, Sharp T, Finocchio M, Moore R, Kipman A, Blake A (2011) Real-Time human pose recognition in parts from single depth images. http://research.microsoft.com/apps/pubs/default.aspx?id=145347
6. Kuipers JB (2002) Quaternions and rotation sequences: a primer with applications to orbits. Princeton University Press, Princeton (Aerospace and Virtual Reality)

7. Welch GF (2009) History: the use of the Kalman filter for human motion tracking in virtual reality. Presence: Teleoper Virtual Environ 18:72–91
8. Richardson DC (2004) Eye-tracking: characteristics and methods. Encyclopedia of biomaterials and biomedical engineering. Stanford University, ZDA
9. Poole A, Ball LJ (2005) Encyclopedia of human computer interaction, chap. Eye tracking in human-computer interaction and usability research: Current status and future prospects. Idea Group, Velika Britanija
10. Duchowski AT (2003) Eye tracking methodology: theory and practice. Springer, London
11. Morimoto C, Koons D, Amir A, Flickner M (2000) Pupil detection and tracking using multiple light sources. Image Vis Comput 18:331–335

Chapter 5
Visual Modality in Virtual Reality

Abstract Sight is perhaps the most important of all human senses, and the visual modality is thus also a key component of virtual reality. This chapter begins with the biological basics of human visual perception, with a particular focus on depth perception. It then explores the basic elements of computer graphics. It describes the basic models used in virtual environments (polygons, implicit and parametric surfaces, constructive solid geometry, solid modeling) as well as the process of rendering a model (via projections and transformations, clipping, determining visible objects, illumination, and finally pixel conversion). The chapter concludes with a description of various visual displays, from two-dimensional liquid crystal and plasma systems to stereoscopic displays such as head-mounted, spatially multiplexed, temporally multiplexed and volumetric displays.

Sight is perhaps the most important of all human senses, and a large part of the brain is dedicated to interpreting information obtained from visible light. The visual modality is thus also a key component of virtual reality; some virtual environments with no visual component do exist, but they are few and far between. This chapter begins with an overview of human visual perception with an emphasis on depth, then continues with the design and displaying of visual elements in virtual reality.

5.1 Human Visual Perception

5.1.1 Light Perception

Figure 5.1 shows the human eye. Visible light from the environment enters the eye via the transparent cornea. Light intensity is controlled by the pupil, which dilates or contracts similarly to a camera shutter and thus limits the amount of light that can enter the eye. Behind the pupil is the lens, which focuses light on the retina. The lens is attached to the ciliary muscle, which controls the thickness of the lens by contracting. This allows objects at different distances from the eye to be seen clearly.

M. Mihelj et al., *Virtual Reality Technology and Applications*, Intelligent Systems, Control and Automation: Science and Engineering 68, DOI: 10.1007/978-94-007-6910-6_5,

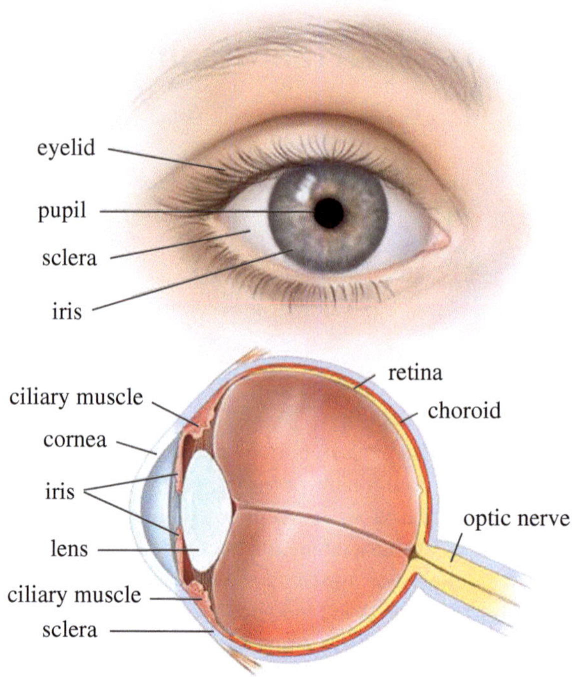

Fig. 5.1 The human eye

The retina contains specialized light-sensing nerve endings called photoreceptors. These can be divided into two groups: cones (sense colors and react more quickly to changes, but are less sensitive to light in general) and rods (do not sense color, but are more sensitive to light and thus allow sight in low-light conditions). The light that falls on these receptors is converted to an electrochemical signal that travels along the optic nerve to a part of the brain called the visual cortex. This cortex converts incoming signals into the actual image that we 'see'.

5.1.2 Color Perception

The human eye senses colors using cones in the retina. Cones are divided into three type, each of which senses different wavelengths of light. The first type senses yellowish light (564–580 nm), the second senses greenish light (534–545 nm) and the third senses bluish light (420–440 nm). We thus, for example, see blue color if the third type of cones is stimulated more than the second. Similarly, we see purple color if the third type of cones is stimulated much more than the second. The range of wavelengths that human eyes can 'see' is between approximately 380 and 700 nm. Light

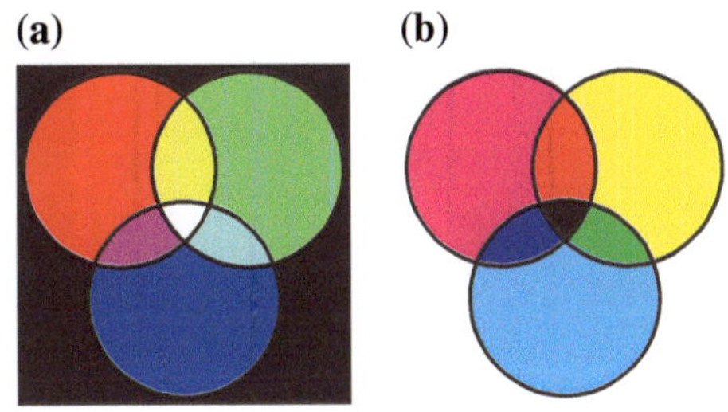

Fig. 5.2 RGB (*left*) and CMYK (*right*) color models

with shorter wavelengths is called ultraviolet while light with longer wavelengths is called infrared.

Since the human eye has three types of cones, computers usually also use a model with three primary colors. Mixing these three colors allows any color to be created. The most frequently used model in computing is the RGB model, which uses red, green and blue as the primary colors (Fig. 5.2a). These colors roughly correspond to the three types of cones. Another popular model is the CMYK model (Cyan-Magenta-Yellow-Black, Fig. 5.2b), which is mostly used for color printing.

5.1.3 Depth Perception

Among other things, our eyes also convey information about depth—the distance between us and particular objects. Depth is especially important in virtual reality, as normal twodimensional visual displays cannot properly incorporate depth into the image. This subsection thus covers the basics of human depth perception, as virtual reality designers need to fool the human senses and create an illusion of depth in the virtual environment.

Humans estimate distances to different objects using so-called depth cues. These are divided into monoscopic and stereoscopic.

Monoscopic depth cues can be seen with only one eye and are also present in twodimensional images. They include:

1. *Occlusions*: Objects in the foreground occlude those in the background (Fig. 5.3).
2. *Shading*: lets us better estimate the shape of an object, but shadows also indicate relative positions of different objects (Fig. 5.3).
3. *Size*: The size of an object can be compared to the size of similar objects we have memorized, thus giving an impression of absolute distance. Comparing the sizes of two similar objects allows us to gauge their relative distance (Fig. 5.3).
4. *Linear perspective*: Parallel lines appear to converge toward a point as they recede into the distance (Fig. 5.3). This is a useful cue with objects consisting of straight lines (e.g. most buildings).

Fig. 5.3 Psychological cues used to perceive depth (linear perspective, shadows, occlusion, texture gradient, horizon, blurring)

5. *Surface texture*: Distant objects have a less sharp texture than close ones since the eye cannot distinguish details at great distances (Fig. 5.3).
6. *Accommodation* (Fig. 5.4) is the process where the lens dilates or contracts, allowing the eye to focus on the object it is viewing. The brain can estimate the distance to the object from the required lens thickness.
7. *Parallax*, the movement of the viewer, allows the distance of the viewed objects to be estimated since distant objects appear to move less in the field of view than nearby ones.
8. *Movement of the viewed object* also allows relative distance to be estimated. As the viewed objects move away from the viewer, they seem to get smaller. When they move closer, they appear larger. Based on this cue, the brain also estimates how much time an approaching object will need to collide with the viewer.

Stereoscopic depth cues combine information from both eyes. The most important cues are:

1. *Convergence* (Fig. 5.4) is the process where the eyes turn toward an object in order to focus on it. The angle of the eyes allows the brain to estimate depth. Convergence always occurs simultaneously with the previously mentioned accommodation.
2. *Stereopsis* (binocular disparity—Fig. 5.5) allows us to estimate depth from the differences between what the left eye sees and what the right eye sees.

The designer of a virtual environment can mix any of these cues and thus create an illusion of virtual depth. Since most of the cues are additive, adding more of them creates a more realistic feeling of depth.

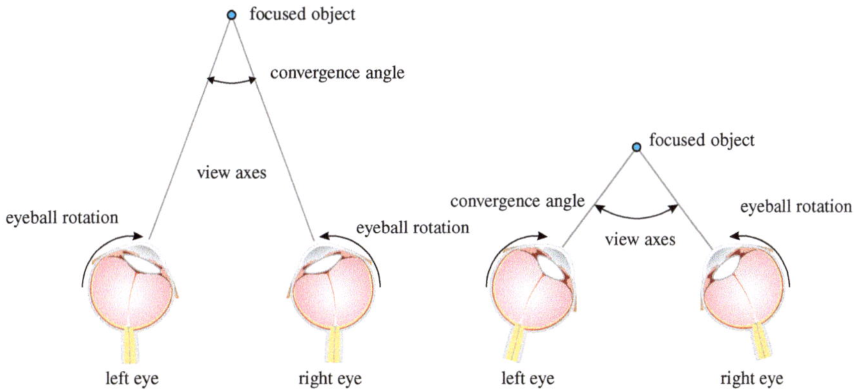

Fig. 5.4 The principle of convergence and accommodation. The eyes turn toward the object, and the eye angle allows the brain to estimate depth (convergence). At the same time, the lens dilate or contract in order to focus on the object (accommodation)

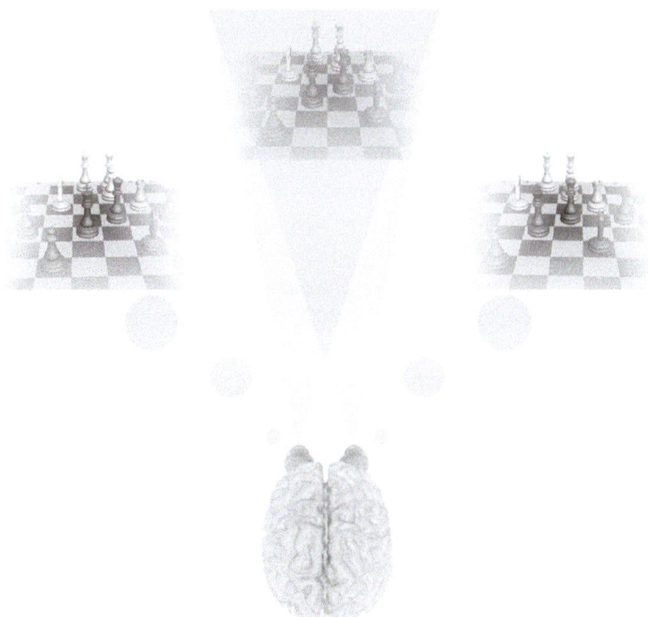

Fig. 5.5 The principle of stereopsis. Due to slightly different viewpoints, the *left* and *right eye* see different images. The differences between the images allow depth to be estimated

5.2 Computer Graphics

While a virtual environment can in principle exist without being seen by anyone, such an environment is not very useful. All elements of the virtual environment should have a defined visual representation that can later be shown to users on a display. The design of these visual representations is in the domain of computer graphics, which uses various specialized hardware and software. This section covers the basics of the field, though several concepts are greatly simplified for easier understanding.

5.2.1 Basic Terms

Computer images are divided into two large categories: raster and vector images.

5.2.1.1 Raster Images

A raster image consists of a two- or three-dimensional grid whose every element has a defined color. An element of a two-dimensional grid is called a *pixel* while an element of a three-dimensional grid is called a *voxel*. The quality of the image is defined by its resolution (number of elements per unit of area or volume) and the number of available colors.

Since raster images require every element of the grid to be defined, the quality of the image cannot be easily increased. Increasing the resolution requires the color of all new image elements to be defined, which is usually done via interpolation. An interpolated image, however, cannot have the same quality as an image that was originally created in a higher resolution (Fig. 5.6). Furthermore, the amount of computer memory required grows very quickly as the resolution increases. Nonetheless, raster images are very suitable for e.g. digital photographs and complex scenes where every part of the image truly needs to be accurately defined.

5.2.1.2 Vector Images

Vector images define individual elements of the image using simple geometric shapes: lines, plane figures and (in three-dimensional images) bodies. By adding and subtracting simple geometric shapes, it is possible to create complex objects called *polygons*. A vector image stored on a computer takes the form of mathematical formulas that describe the image. Unlike raster images, this allows the resolution of the image to be increased easily without increasing the amount of required computer memory (Fig. 5.6).

Fig. 5.6 Increasing the resolution of vector (*top left*) and raster images (*top right*)

5.2.2 Modeling Virtual Environments

In order to generate a visual scene, a computer needs an appropriate model of all the objects that describe the scene. We have already described methods of defining the position and physical properties of objects, which will not be repeated here. This section instead focuses on describing the shape and color of each object. It is possible to model either only the exterior surfaces of an object (*boundary representations*) or the entire object (*solid modeling*).

5.2.2.1 Boundary Representations

Boundary representation methods are suitable for nontransparent objects where information about the interior is unnecessary. They are generally based on polygons, implicit/parametric surfaces or constructive solid geometry.

Polygons

Polygons are the simplest modeling method and consist of plane figures with at least three straight edges. Though any number of edges can in principle be used, polygons with three or four edges are the most common in practice. Triangular polygons have three major advantages: they are always planar, they are always convex, and any

Fig. 5.7 The model of a sphere consisting of a *low* (*left*) and *high* (*right*) number of triangular polygons. Although each polygon is a plane figure, a sufficiently large amount of polygons allows curved surfaces to be modeled with any desired accuracy

complex polygon can be cut up into multiple triangular ones. Any curved surface can thus be modeled with any desired accuracy by using a sufficiently large number of triangular polygons. The main limitation is the amount of memory required to store the polygons, which increases rapidly with accuracy. A polygonal object model has the appearance of a wire mesh (Fig. 5.7).

A polygon is completely described with not only the positions of its vertices, but also its color, textures and surface parameters. Object representations are frequently simplified by grouping polygons into sets that represent specific objects (e.g. a chair or table). Grouping allows an object to be easily moved as a whole without needing to move e.g. an individual table leg or even individual polygons.

Implicit and Parametric Surfaces

Implicit and parametric surface modeling methods describe curves with mathematical equations. They allow certain curved objects such as spheres to be described with far less information than in the case of polygons. Parametric surface modeling describes each curve as a function of one or more parameters while implicit surface modeling describes each function as an implicit function of multiple variables. Due to their mathematical nature, these descriptions are perfectly accurate: the same amount of information allows any desired resolution to be obtained.

For example, the parametric definition of a sphere is:

$$\begin{aligned} x(u, v) &= r\cos(u)\cos(v) \\ y(u, v) &= r\cos(u)\sin(v) \\ z(u, v) &= r\sin(u). \end{aligned} \tag{5.1}$$

where r is the radius of the sphere, u runs from $-\pi/2$ to $\pi/2$, and v runs from 0 to 2π. All possible (x, y, z) points lie on the surface of the sphere.

Similarly, the implicit definition of a sphere is:

$$x^2 + y^2 + z^2 = r^2 \tag{5.2}$$

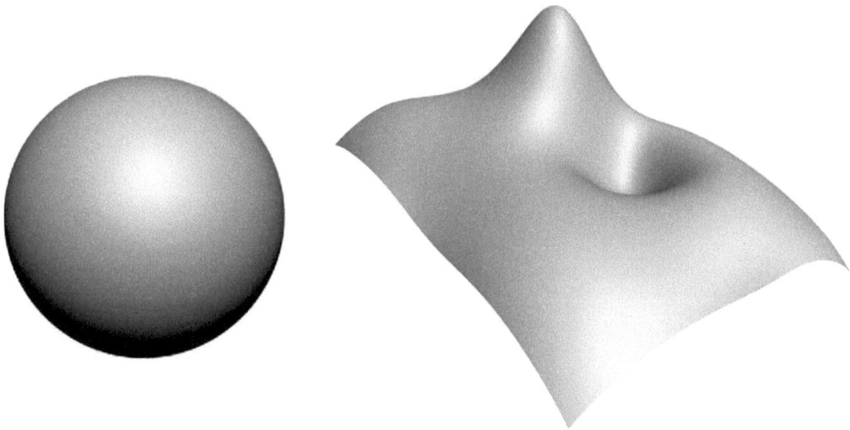

Fig. 5.8 Example of an implicitly modeled sphere surface (*left*) and a surface modeled with splines (*right*)

All (x, y, z) points that satisfy this requirement lie on the surface of the sphere with the radius r (Fig. 5.8 left).

A sphere can thus be perfectly described with a single equation. If we wanted to use polygons instead, we would need to save thousands and thousands of polygons into the computer, and the modeled sphere would still not be perfectly round. Higher-order parametric and implicit equations allow even extremely complex curves to be modeled. Among these higher-order equations, *splines* deserve special mention. They are curves created from several polynoms 'spliced' together piece by piece. A simple two-dimensional spline, for instance, is:

$$y = \begin{cases} f(x) & \Leftarrow \quad x < 0 \\ g(x) & \Leftarrow \quad 0 \leq x < 1 \\ h(x) & \Leftarrow \quad x \geq 1 \end{cases}$$
$$f(0) = g(0), \ g(1) = h(1). \tag{5.3}$$

A spline is always continuous—each piece fits together with the adjoining one. Their first or even second derivatives are usually also continuous, ensuring a smooth curve. The above example would require the following to ensure a continuous first derivative:

$$f'(0) = g'(0), \ g'(1) = h'(1). \tag{5.4}$$

Splines thus allow different curves to be approximated. The concept can be easily expanded to three dimensions: pieces of a surface (described with simple polynoms) are 'stitched' together at the edges, thus creating a more complex surface (Fig. 5.8 right).

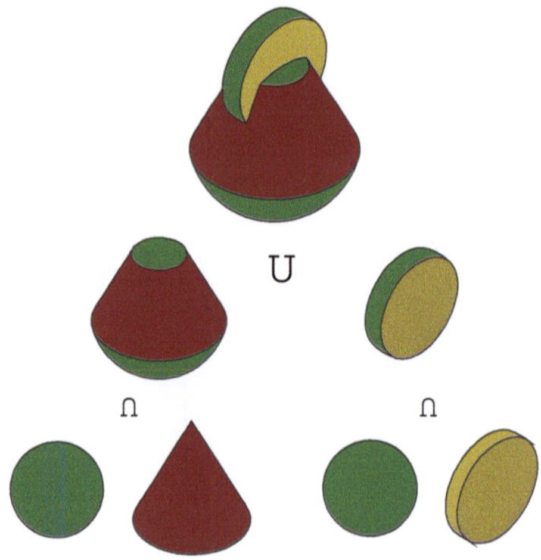

Fig. 5.9 An example of constructive solid geometry

Constructive Solid Geometry

Constructive solid geometry is based on combining simple geometric bodies. The most common basic bodies (also called primitives) are the cube, triangular prism, sphere, cylinder, cone and torus. Complex bodies created with this method can be defined as a sequence of mathematical and logical operations (conjunction, disjunction, difference, scaling. . .) performed on the primitives (Fig. 5.9).

Similarly to implicit and parametric surface modeling, constructive solid geometry allows complex shapes to be accurately described with a small amount of information compared to polygons. Additionally, it is much more intuitive than implicit or parametric surface modeling. Its main disadvantage is that the same body can be created with different primitives and operations.

So Why Polygons?

Since implicit/parametric surface modeling and constructive solid geometry can accurately model smooth curves with a very small amount of information, we might wonder why anyone would want to use polygons at all.

In fact, polygons are one of the basic building blocks of computer graphics. This is because of their speed: most modern graphics cards are optimized to display polygons. Graphics cards can thus display virtual environments created using a large number of polygons and refresh them with a frequency above 50 Hz. On the other hand, implicit/parametric surface modeling and constructive solid geometry use ray tracing to create pixels. It is difficult to achieve refresh rates above 10 Hz using ray tracing and today's hardware. Furthermore, objects modeled with implicit or parametric surfaces can be (with some data loss) relatively easily converted to polygons.

Conversion of constructive solid geometry to polygons is, unfortunately, much more difficult.

5.2.2.2 Solid Modeling

Boundary representations are inappropriate for transparent objects where the interior also needs to be modeled. This is especially true for spaces that contain a partially transparent substance of varying density (e.g. mist, magnetic resonance images). Solid (non-geometric) modeling comprises volumetric modeling and particle systems.

Volumetric modeling is suitable for partially transparent objects and is frequently used to display medical, seismic or other research data. It is based on ray tracing: light rays, which obey the laws of physics, change their properties upon being reflected off virtual objects or passing through partially transparent materials.

Particle systems are commonly used to display complex flows in visual scenes. A large number of particles are generated in the environment and then move according to predefined physical laws (acceleration, gravity, reflection). Their initial position and velocity are defined only roughly (e.g. with the mean value and standard deviation). The movement of such particles allows simulation of very complex phenomena such as fire, smoke and large groups of people.

5.2.2.3 Scene Graph

A scene graph is a tool that allows accurate and flexible representation of the virtual environment's hierarchy. Its structure defines the relative position and orientation of all objects in the virtual environment as well as other object features such as color and texture. It is thus possible to change a part of the virtual world with a single change to the scene graph. An example of a scene graph is shown in Fig. 5.10. In this case, it is possible to move (open) a drawer and its contents by changing a single coordinate system. Similarly, it is possible to move a book to the drawer with a single operation.

5.2.3 Rendering

Rendering is the process of converting a model of a virtual environment to an actual image. The model is usually three-dimensional and can have been made using polygons, constructive solid geometry or any other aforementioned method. The output image is two-dimensional and can be displayed on the screen.

Several steps are required to create the image. We must first define the point from which the virtual world is viewed. This point will henceforth be referred to as the 'camera'. We must then determine the part of the environment that we are

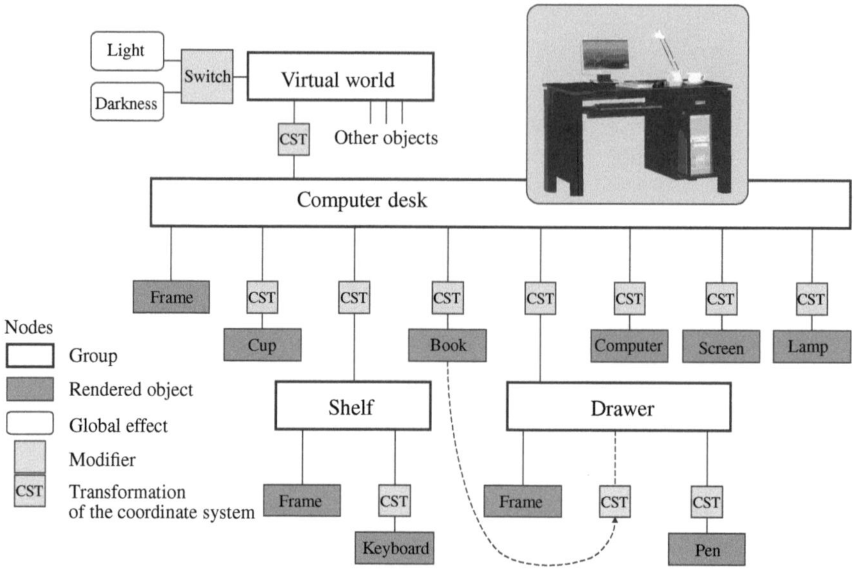

Fig. 5.10 The scene graph allows related objects to be combined so that their parameters can be defined more easily

viewing, as the rest does not need to be rendered. If we want to include shading and different levels of light in the image, we need to simulate lighting in the environment. Finally, the virtual environment visible from the camera must be transformed into a two-dimensional projection and drawn on the screen.

5.2.3.1 Projections and Transformations of the Virtual Environment

At the start of rendering, the virtual environment must be transformed from the global coordinate system to the camera's coordinate system. The position and orientation of the camera in the virtual environment must first be determined. Once they are known, we can use equations given in Chap. 2 to easily determine the transformation matrix from the global coordinate system to the camera's coordinate system. This results in a three-dimensional environment seen from the camera's viewpoint.

Displaying the virtual environment on a screen requires a two-dimensional image of the environment as seen from the perspective of the camera. Thus, a projection from three to two dimensions is also required. Such projections are divided into parallel and perspective projections (Fig. 5.11). Parallel projections are mostly used in technical drawing and assume that the camera (center of projection) is located an infinite distance from the object. Lines that are parallel in three dimensions thus also remain parallel in a two-dimensional image. Perspective projections, on the other hand, assume that the camera is near the objects. Lines that are parallel in

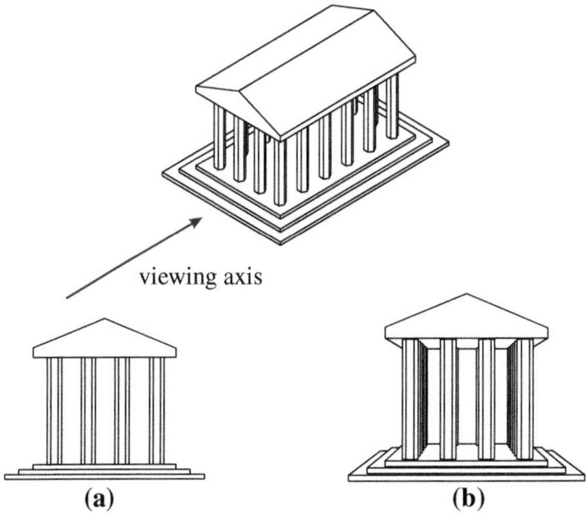

Fig. 5.11 Parallel projection (*left*) and perspective projection (*right*)

a three-dimensional environment thus are not necessarily also parallel on a two-dimensional image. The size of the object decreases as the distance from the camera increases. Perspective projections are more common than parallel ones in virtual reality since they create a realistic image and illusion of depth.

Virtual reality often uses stereoscopic displays, which create an illusion of depth by displaying slightly different images to the left and right eyes. In this case, we define two cameras that view the environment from slightly different positions, then perform the projection and other steps separately for each camera.

5.2.3.2 Clipping

Objects outside the camera's field of view do not need to be rendered by the computer. Clipping is thus the analytical process of determining which parts of virtual objects lie outside the camera's field of view. This can be done using several different algorithms. Lines are generally trimmed with the Cohen-Sutherland or Cyrus-Beck algorithm while polygons are clipped with the Sutherland-Hodgeman, Weiler-Atherton or Vatti algorithm. Their mathematical details are beyond the scope of the book, though it is important to be aware of the general problem.

5.2.3.3 Determining Visible Objects

A perspective projection transforms a three-dimensional body to a two-dimensional figure. However, if there are several objects in the virtual environment, it is necessary

Fig. 5.12 Consecutive steps of the painter's algorithm

to determine which objects are obscured by others. The obscured objects then do not need to be drawn on the screen. Several algorithms exist to solve this challenge. The simplest is the so-called *painter's algorithm*, which sorts polygons according to their distance from the camera and then draws them in order from the most distant to the nearest one (Fig. 5.12). It thus paints over distant objects with nearby ones. This approach is not very efficient, as it nonetheless draws all obscured polygons. This weakness can be solved with the *reverse painter's algorithm*, which draws polygons from the nearest to the most distant. In each step, it colors only those pixels that have not yet been colored.

Both the standard and reverse painter's algorithms have several weaknesses and have thus been replaced in practice by the *depth buffer*, which is also called the Z-buffer. This buffer solves the visibility problem for each pixel separately. It takes the form of a two-dimensional matrix where each element represents a single pixel. When the computer renders an object in a particular pixel, it saves the distance (depth) of this object in the corresponding element of the buffer. Each pixel of the object can have its own depth. If the computer later tries to render a new object in the same pixel, it first compares the distance of the new object to the value in the depth buffer. If the new object's distance is smaller, the computer draws the new object in the pixel and updates the value of the buffer.

An alternative method worth mentioning is *ray casting*, which uses an entirely different principle than the above algorithms. It creates several rays (one for each pixel) which travel from the camera in different directions. The first object hit by a ray is then drawn by the computer in the ray's corresponding pixel. This method is relatively slow compared to the depth buffer method, but has certain other advantages. It represents a simplified ray tracing algorithm which we will examine in the lighting and shading subsection.

5.2.3.4 Light and Shadows

So far, we've defined the point from which we view the virtual environment and the objects that we can see. However, in the real world we would not be able to see these objects if they were not illuminated by the sun or another source of light. Thus, different objects are more or less visible depending on the distance from

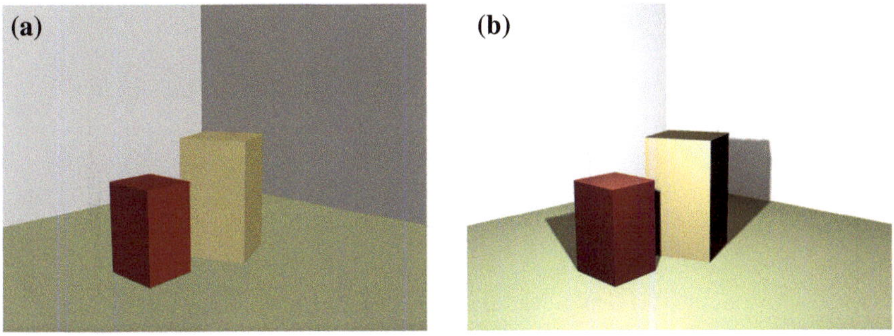

Fig. 5.13 Unlit (*left*) and lit (*right*) object

the light source and its intensity. The object's properties such as color and albedo also influence its lighting, and each object casts its shadow onto other objects. In virtual reality, it is in principle possible to ignore realistic lighting and assume that all objects are perfectly illuminated. Adding shadows may not really convey much practical information to the viewer, but it can vastly improve the feeling of realism (Fig. 5.13).

It can be very computationally demanding to take into account all possible sources of light (with different intensities, colors, positions...) and all possible occlusions between objects. Thus, two simplifications are often used: illumination without any light sources and local illumination.

Illumination Without Light Sources

The simplest implementation of lighting in virtual reality does not use any light sources, but simply changes the color or brightness of different objects depending on their distance from the camera or other reference point. Objects farther from the camera are thus darker, simulating poor visibility of distant objects. The method is computationally extremely simple and is used in applications such as landscape modeling or medical image displays.

Local Illumination

Local (also called direct) illumination models allow light to travel from a light source to an object and be reflected from the object, but the reflected light cannot then hit a second object. This allows, for example, the sides of an object that are facing toward a light source to be illuminated more than sides facing away from the light source. However, local illumination cannot model phenomena such as shadows.

There are three types of light in local illumination:

Fig. 5.14 Diffuse lighting (*left*) and specular lighting (*right*)

1. *Ambient lighting* comes from all directions and evenly illuminates the entire object. Two neighboring sides of the same object are illuminated identically, so no border is visible between them.
2. *Diffuse lighting* comes from a specific direction, but is reflected equally in all directions when it hits an object and thus doesn't depend on the position of the camera. It appears with rough surfaces such as chalk (Fig. 5.14).
3. *Specular lighting* comes from a specific direction and is reflected in another specific direction when hitting an object (Fig. 5.14). The amount of visible specular light thus also changes with camera position. It appears with smooth or shiny surfaces such as mirrors or metal.

Each object has its own separate reflection coefficients for ambient, diffuse and specular light reflection.

A local illumination model can use either ambient lighting, ambient lighting in combination with diffuse lighting, or all three lighting types. The most commonly used model is Phong illumination, but its mathematical details are beyond the scope of this book.

Global Illumination

Global illumination models take into account not only the light that reaches an object directly from the light source, but also light reflected onto the object from other objects. This gives the most realistic image of the environment, but also significantly increases computational demands.

The best-known global illumination model is *ray tracing*, where light 'rays' are sent from the camera in different directions. Their progress through the virtual environment is then traced. When a ray falls on an object, it can be reflected and thus travel on. If the ray arrives to an object directly from the camera, reflection can be diffuse or specular. If the ray was previously reflected by another object, reflection can only be specular. Each ray travels until it is reflected a certain amount of times or it covers a predefined distance without any reflections occurring. The color and brightness of each point thus depends on all rays reflected from it.

Ray tracing is very computationally demanding, so some simple applications instead use the ray casting algorithm. Ray casting is very similar to ray tracing, except that a ray stops the first time it hits an object.

An alternative to ray tracing is the *radiosity* model, which was originally used for heat transfer models and only incorporates diffuse lighting. The algorithm is based on finite element modeling and thus requires virtual objects to be first divided into smaller surfaces (e.g. polygons). For each of these surfaces, the algorithm calculates the incoming light energy (which comes from light sources and all other surfaces in the environment) and the outgoing light energy, which depends on the reflection coefficient. The calculations are usually iterative: the first step calculates the energy coming from light sources to all surfaces, the second step calculates the energy transferred in the first reflections, and so on.

Neither ray tracing nor radiosity are perfect illumination models, as ray tracing principally models specular reflection while radiosity only models diffuse reflection. If a better image is desired, both approaches can be combined: radiosity is first used to calculate diffuse reflection, and diffuse light is then taken into account for ray tracing. This creates a more realistic image which is of course still not perfect (for example, it does not include light that is first reflected specularly and then diffusely). Of course, there are also other, less popular global illumination methods such as photon mapping.

5.2.3.5 Conversion to Pixels

The last rendering step is to convert the virtual environment into color pixels that can be directly shown on the screen. This conversion is divided into multiple algorithms that are responsible for drawing different components of the environment. They include:

1. *Line drawing*: The algorithm uses the environment model to obtain the start and end points of all straight lines, then determines the intermediate pixels that need to be drawn. The problem is that a pixel is approximated as a small square. Combining multiple such small squares only allows horizontal or vertical lines to be drawn, not diagonal ones. The algorithm must thus draw a diagonal line as 'steps' that approximate a line such that the error is smallest (Fig. 5.15). Examples

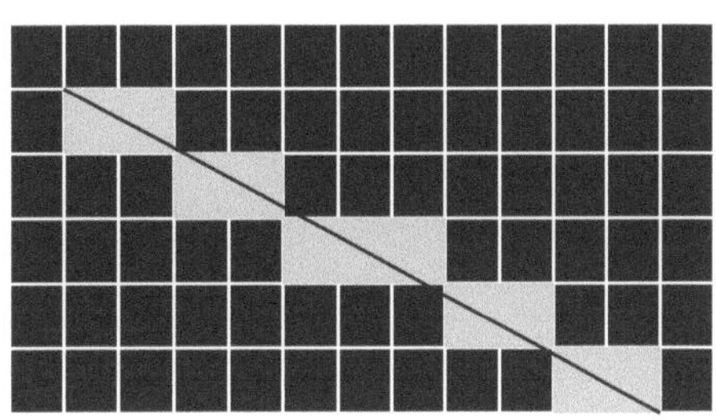

Fig. 5.15 Bresenham's line algorithm draws a line as a series of steps that approximate a line with the smallest possible error

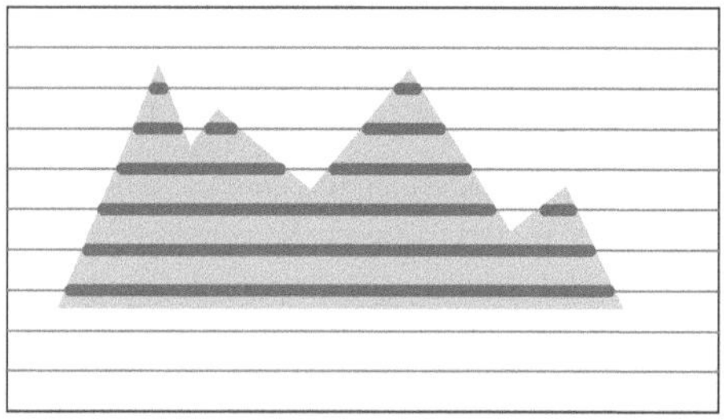

Fig. 5.16 Polygon scan conversion, which "colors" the pixels inside a polygon. For every row of pixels, a *horizontal line* is drawn. The line is divided into sections, with every intersection between the line and polygon marking the beginning of a new section. Every second section is then colored

of line drawing algorithms are the digital differential analyzer and Bresenham's line algorithm.

2. *Circle drawing*: Drawing circles presents a similar challenge as drawing diagonal lines: the smoothness of the circle needs to be approximated with a multitude of small squares. When drawing, parametric or implicit circle equations can be used to speed up the procedure. Since a circle is symmetric, the equations only need to be evaluated for one eighth of the circle, and symmetry can be used to obtain the other points. For instance, if the center of the circle is at (x_0, y_0) and the point $(x_0 + a, y_0 + b)$ lies on the circle, it can immediately be determined that points $(x_0 \pm a, y_0 \pm b)$ and $(x_0 \pm b, y_0 \pm a)$ also lie on the circle. The most popular circle drawing algorithm is Bresenham's circle algorithm.
3. *Polygon drawing*: It was previously mentioned that polygons are a basic building block of computer graphics. There are thus several algorithms that specialize in rasterizing different types of polygons (triangular, convex etc.). The best-known algorithm is the so-called 'polygon scan conversion', which colors all pixels inside a polygon. However, it first needs to find all pixels inside the polygon, which can be very challenging for concave polygons. For every row of pixels, a horizontal line is drawn. The line is divided into sections, with every intersection between the line and polygon marking the beginning of a new section. Every second section is then colored (Fig. 5.16). The algorithm is fast and effective, and can be tweaked to speed it up even further.

A special example of conversion into pixels is the previously mentioned ray tracing. It is used for implicit and parametric surface modeling as well as constructive solid geometry. The algorithm sends light 'rays' from the camera in different directions, then follows their paths. When the ray hits an object, it can be reflected and thus travel on. Each ray travels until it reaches a maximum number of reflections or travels a certain distance without any reflection. The color and brightness of each pixel thus depends on all the rays reflected from it. The algorithm represents an effective alternative pixel conversion method and can create a more realistic image, but is relatively slow and thus far less popular.

5.3 Visual Displays

Visual displays are defined as the hardware that presents the created visual image to the user.

5.3.1 General Properties of Visual Displays

This section covers multiple displays that differ in a large number of properties. These properties define the quality of the image and thus the quality of the virtual reality experience or may have a major impact on the display's practical usefulness. For example, some stationary displays need a large amount of space while head-mounted displays usually require little space.

Color—most displays use the trichromatic system where three primary colors (red, green and blue) allow the entire color palette to be generated. The trichromatic makes sense since the human eye has three types of color-detecting cones. For augmented reality, monochromatic displays are sometimes preferable for their higher brightness and contrast.

Spatial resolution is usually given as the number of pixels per unit of length. The size of the display and the number of pixels affect the image quality. The distance between the eyes and the display must also be taken into account, as a smaller distance requires a higher pixel density.

Contrast is the difference between bright and dark regions of the image. A higher contrast makes it easier to separate different components of the image.

Focus distance is the virtual distance between the image and the user's eyes. Current technology places all objects in the scene at the same focus distance regardless of their virtual distance from the observer. The disparity between different visual depth cues causes problems with depth perception, which can lead to headaches and sickness.

Transparency—the display can hide the real world from the user or allow it to be viewed. Stationary screens and desktop displays cannot hide the real world and are thus 'transparent'. Head-mounted displays, on the other hand, are usually nontransparent and prevent the wearer from seeing the real environment. Nontransparency can affect safety (real objects cannot be seen) and the possibility of cooperation with others (other people cannot be seen).

Occlusion—when using stationary displays, objects from the real world (such as the user's hand) can occlude virtual objects. This is most problematic when a virtual object comes between the user's eyes and a real object. In this case, the virtual object should occlude the real one, but this does not occur. The problem is more easily solved using head-mounted displays.

Field of view represents the width of the currently visible world. The human field of view is approximately 200° with 120° of binocular overlap. The overlap area is the most important, so displays with a 120° field of view cover an acceptable portion of human view. The field of view of a head-mounted display, for instance, defines how much of the world the user can see without having to turn his/her head.

Acceptable graphical delay is prominent in displays that change the view of the virtual environment in response to the user's movement. There is a certain delay between the user's movement and the corresponding change on the display, which can make the user uncomfortable or even sick. The delay must thus be as small as possible. The acceptable graphical delay in augmented reality is even smaller than the acceptable delay in normal virtual reality, as a large delay means that the real and virtual worlds are no longer synchronized.

Temporal resolution represents the frequency with which the displayed image is refreshed. It significantly affects the user's experience of the virtual world. A frequency above 30 Hz is acceptable while frequencies under 15 Hz lead to discomfort.

User mobility can affect both virtual presence and the usefulness of the virtual reality. Most displays limit the user's mobility as a consequence of user movement sensors, electrical connections or stationary displays.

User movement sensing—advanced displays can measure the user's movement and adapt the displayed image accordingly.

Compatibility with displays for other senses—head-mounted displays are entirely compatible with headphones, but speakers are preferred with stationary screens. Haptic displays present a special challenge, as their limited workspace also creates limitations for other (e.g. visual) displays.

5.3.2 Two-Dimensional Displays

5.3.2.1 Liquid Crystal Displays

Liquid crystals only allow certain wavelengths of light to pass through their unique structure. They thus represent a polarizing filter. Their structure changes in the presence of an electrical field, so the crystals can be made transparent to different wavelengths of light by modulating the electric field. This principle is exploited by liquid crystal displays.

A normal liquid crystal display consists of six layers. The layer farthest away from the viewer is the light source, which can be passive (a mirror reflecting light from the environment) or active (a lamp). The other five layers form a 'sandwich': the two outer layers are polarizing filters (vertical and horizontal), the two middle layers are glass plates with electrodes attached, and the inner layer are liquid crystals. Light travels from the source through the first filter and becomes polarized, then passes through liquid crystals which may allow it to pass completely or only partially. Each pixel of the screen has its own liquid crystal cell, so each pixel's brightness can be modulated individually. A color screen is made similarly except that each pixel consists of three subpixels with different color filters.

5.3.2.2 Plasma Displays

A plasma display consists of millions of small fluorescent lamps. Each lamp contains a mixture of noble gases and mercury. The lamp's inner surface is also covered with fluorescent substances. When a charge is applied to the lamp, the mercury turns to gas while the noble gasses are partially ionized and form the plasma that gives the display its name. The electrons in the mixture collide with mercury particles, increasing the mercury's energy. The mercury then emits energy in the form of ultraviolet light. When this light hits the fluorescent substance on the wall of the lamp, the substance emits heat and visible light that can be seen by the user. An image is formed by manipulating the brightness of the individual lamps. Similarly to liquid crystal displays, a color plasma display is made by having each pixel's lamp consists of three colored sublamps.

The plasma displays of today are heavier and use more energy than liquid crystal displays, but have a faster response time and better contrast.

5.3.2.3 Projectors

In a projection display, the screen itself does not create an image. Instead, the image is projected onto the screen by a separate projector that can be either in front of or behind the screen. The screen is usually much larger than with typical desktop computers and thus covers a larger portion of the user's field of view. Most projectors in virtual reality systems are high-quality and can be based on various technologies (liquid crystals, micromirrors, laser diodes). Multiple projectors and screens can also be used to surround the user and thus increase the field of view.

5.3.3 Stereoscopic Displays

As previously mentioned, human depth perception is partially based on the differences between what the left and right eyes see. Since the object is seen from two viewpoints, the differences can be used to estimate depth. This is not possible with classic visual displays since both eyes see the same image. Stereoscopic displays create an artificial impression of depth by showing a slightly different image to each eye.

The first such display was patented in 1838 by Charles Wheatstone. It consisted of two mirrors and two images (Fig. 5.17). The images showed the same environment, but from two slightly different viewpoints. The mirrors were placed closely together in such a way that the user's left eye saw the left mirror and the right eye saw the right mirror. Each mirror showed one image, thus creating the illusion of depth—each eye saw a slightly different image.

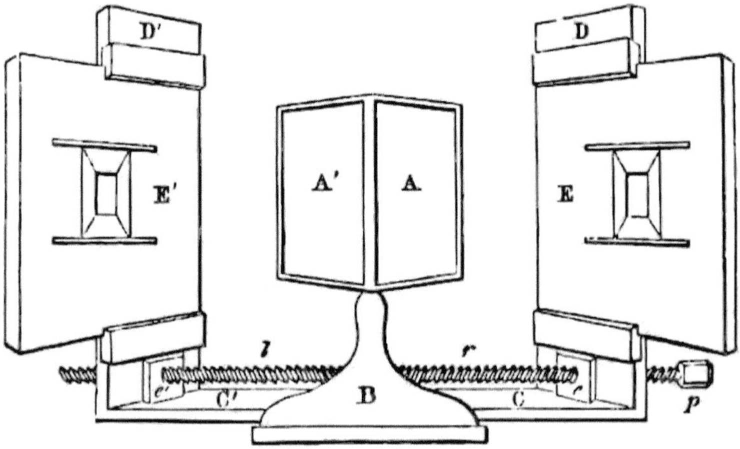

Fig. 5.17 The Wheatstone stereoscope (1838): To view images using the stereoscope, the viewer places his/her eyes directly in front of the mirrors. The mirrors are placed at a 90° angle. The observed images are placed on the attachments at each side of the viewer and are thus spatially separated

Fig. 5.18 Stereoscopic displays show slightly different images to the *left* and *right eyes*. The distance between the object's location on the *left image* and its location on the *right image* is called parallax

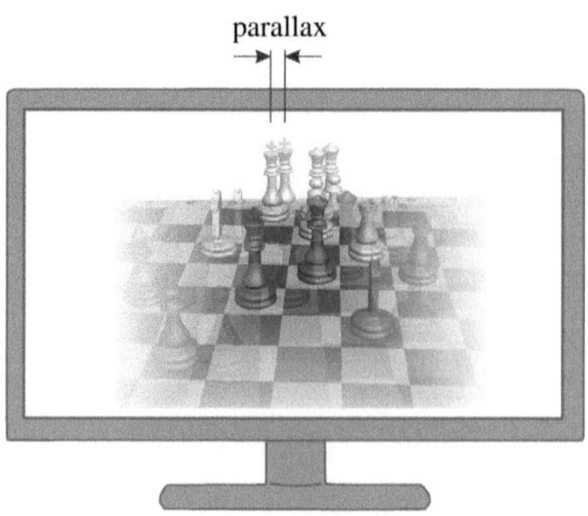

5.3.3.1 Parallax

The difference between an object's location in the image for the left eye and its location in the image for the right eye is called *parallax* (Fig. 5.18). It depends not only on the display, but also on the viewer's position and the distance between the viewer's eyes. It is divided into horizontal and vertical parallax. Horizontal parallax is necessary for the illusion of depth while vertical parallax is always an error in programming or displaying the virtual environment. After all, since the left and right eyes are at the same height, the two images should also be at the same height.

In the conditions shown in Fig. 5.19, horizontal parallax can be calculated with the following equations:

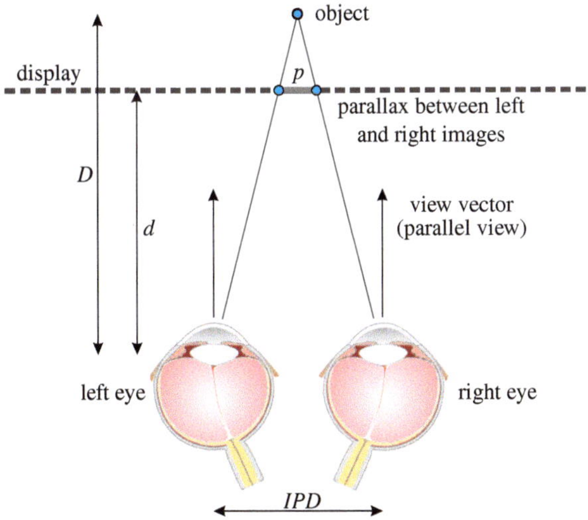

Fig. 5.19 Calculating parallax using geometric rules

$$\begin{aligned} \frac{IPD}{D} &= \frac{p}{D-d} \\ p &= \frac{D-d}{D} IPD, \end{aligned} \tag{5.5}$$

where p is the parallax, IPD is the interpupillary distance, d is the distance between the viewer and display (as well as the point that the eyes focus on for accommodation), and D is the distance between the viewer and the point that the eyes focus on with convergence.

We thus define four separate types of parallax (Fig. 5.20):

1. *Zero parallax* occurs when $D = d$. The object is seen on the screen. Convergence and accommodation focus on the same point, so the object can be viewed without problems.
2. *Positive parallax* occurs when $D > d$. The viewer thus has the impression that the object is inside or behind the screen.
3. *Negative parallax* occurs when $D < d$. The viewer thus has the impression that the object is 'floating' in front of the screen.
4. *Divergent parallax* occurs when the eyes focus on different points due to improper display settings.

Convergence/Accommodation Disparity

Divergent parallax is very unpleasant for the viewer, so we try to avoid it in practice. However, positive and negative parallax can also be unpleasant. The eyes use

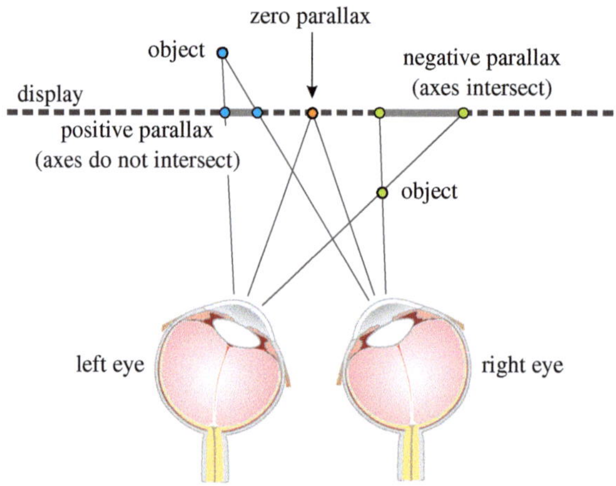

Fig. 5.20 Examples of positive, negative and zero parallax

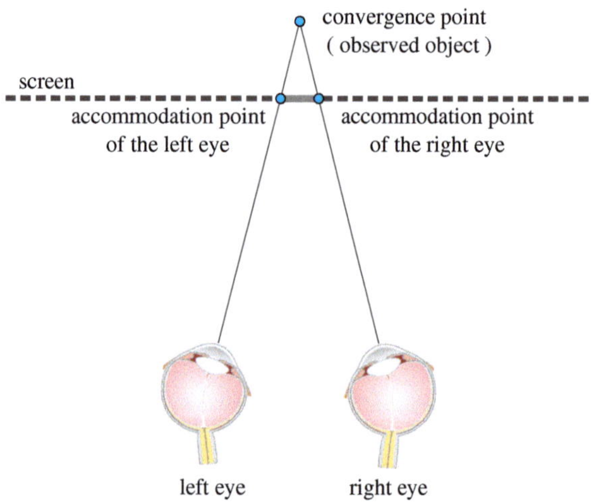

Fig. 5.21 The problem of convergence/accommodation disparity

convergence (turning the eyeballs) to focus on a point in front of or behind the screen (depending on the position of the observed object), but accommodation (lens dilation and contraction) always focuses on the screen itself. The disparity between convergence and accommodation is shown in Fig. 5.21. When choosing a stereoscopic display, it is thus important to find a compromise between realistic depth and pleasantness for the viewer.

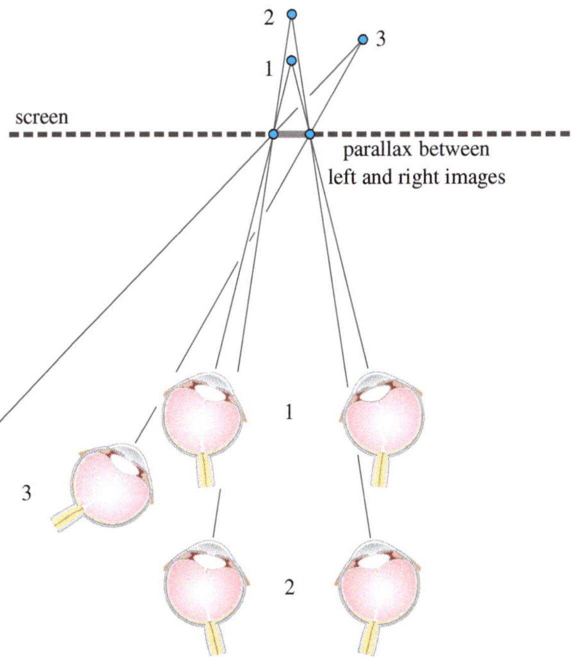

Fig. 5.22 The effect of user viewpoint changes on the perceived object position due to parallax

Viewpoint Changes

Let's take a look at how changing the viewpoint in the presence of parallax affects the perception of the displayed object's position. Figure 5.22 shows the conditions for three people looking at the screen from three different positions. Both the virtual depth of the object and its virtual position in the plane of the screen change with the viewer's position. If the viewer thus moves his/her head in any direction, the object on the screen will also appear to move.

Of course, even large head movements don't allow the viewer to look *behind* an object unless a head movement tracking method is used (for example, viewing a movie where images for the left and right eyes are generated independently of the viewer's position). Viewpoint changes are shown in Fig. 5.23. When a real object is viewed (Fig. 5.23 top), changing head position also changes the perspective. This is only possible in virtual reality if head movements are tracked and new views of the virtual environment are generated accordingly (Fig. 5.23 middle). If head movements are not tracked, changing the viewpoint only slightly changes the object's position on the screen due to parallax (Fig. 5.23 bottom).

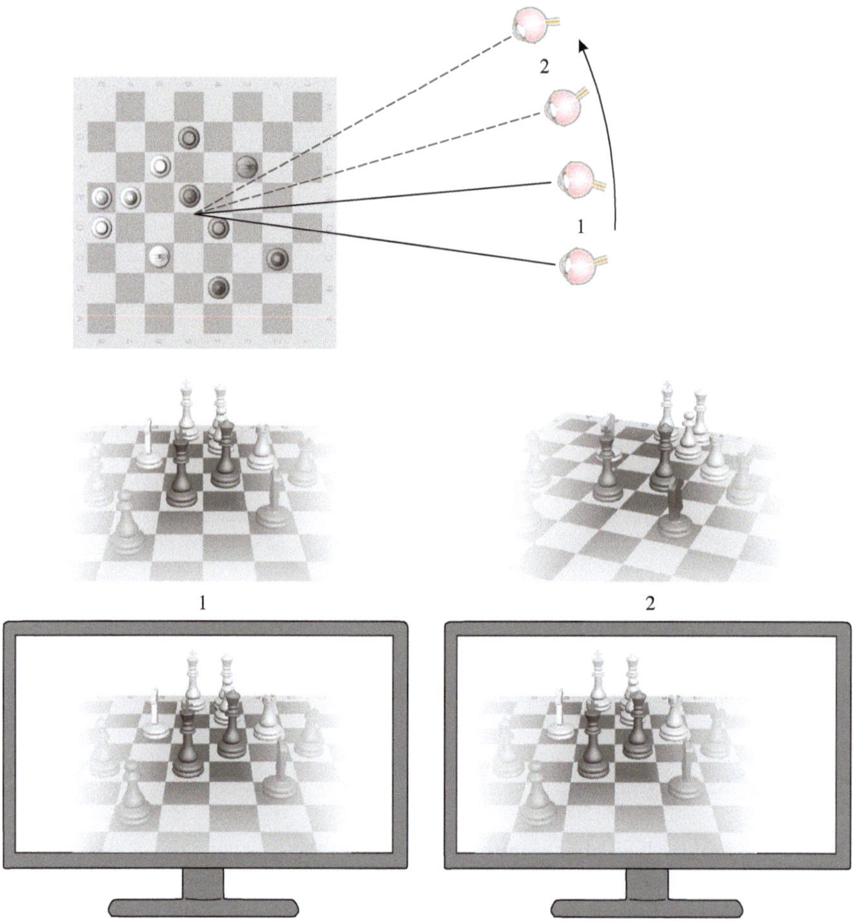

Fig. 5.23 The effect on viewpoint changes on the view of a real (*top*) and virtual object (*bottom*)

5.3.3.2 Head-Mounted Displays

Head-mounted displays have the appearance of glasses whose lenses have been replaced with two screens (one per eye) showing the virtual environment (Fig. 5.24). The screens are usually small and light, so all head-mounted displays are mobile and move together with the user. The illusion of depth is created by having each display show the virtual environment from a different viewpoint. The display follows the user's head movements and changes the view of the virtual environment accordingly.

A frequent weakness of head-mounted displays is the delay between head movement and viewpoint change, which can lead to simulator sickness. The viewing angle of typical head-mounted displays is generally limited while the resolution is usually very high due to the close proximity of the eyes to the screen.

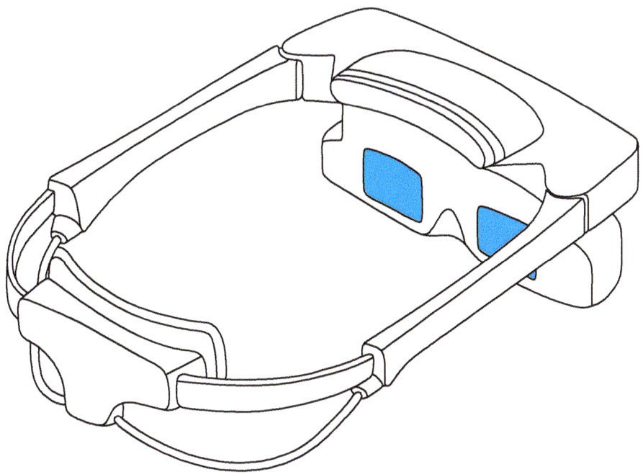

Fig. 5.24 Head-mounted display

The individual screens of the head-mounted display can be transparent or opaque to light. Opaque displays block the view of the real world, so everything that the wearer sees needs to be generated artificially—including the wearer him/herself. Transparent displays are used mainly in augmented reality where the virtual and real environments need to be intelligently combined. They can be created using lenses, mirrors and semitransparent mirrors. Alternatively, they can be created using an additional videocamera that records an image of the real world and superimposes the virtual environment on it. Since the real world is part of the environment in this case, the limitations of the real world affect the limitations of the virtual world. We thus cannot, for instance, change the gravity in the virtual environment.

A special case of head-mounted displays are retinal projectors. Instead of screens, they have two projectors that display the image directly onto the retina. The user thus sees an image 'floating' in the air in front of him/her. Such an image has a higher quality and is less affected by eye imperfections (e.g. farsightedness). Unfortunately, such displays are currently still in the development phase and not yet suitable for mass use.

5.3.3.3 Spatially and Temporally Multiplexed Displays

Unlike head-mounted displays, which use two screens (one per eye), these displays use only one screen. However, each eye obtains different information from the screen. This can be achieved in two ways:

1. The image on the screen can be spatially multiplexed and thus simultaneously contain information for both eyes. Polarized glasses are used to remove part of this information so that each eye only sees the image meant for that eye. The advantage of such displays over head-mounted displays is that the same image

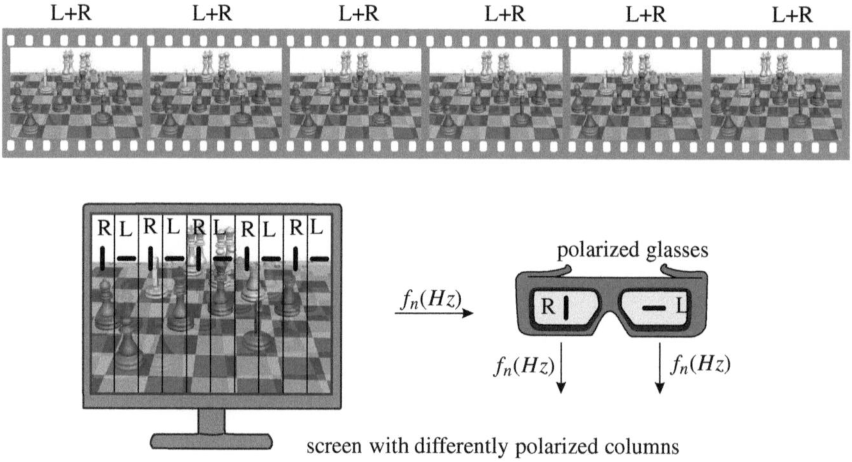

Fig. 5.25 Spatial multiplexing on an LCD screen

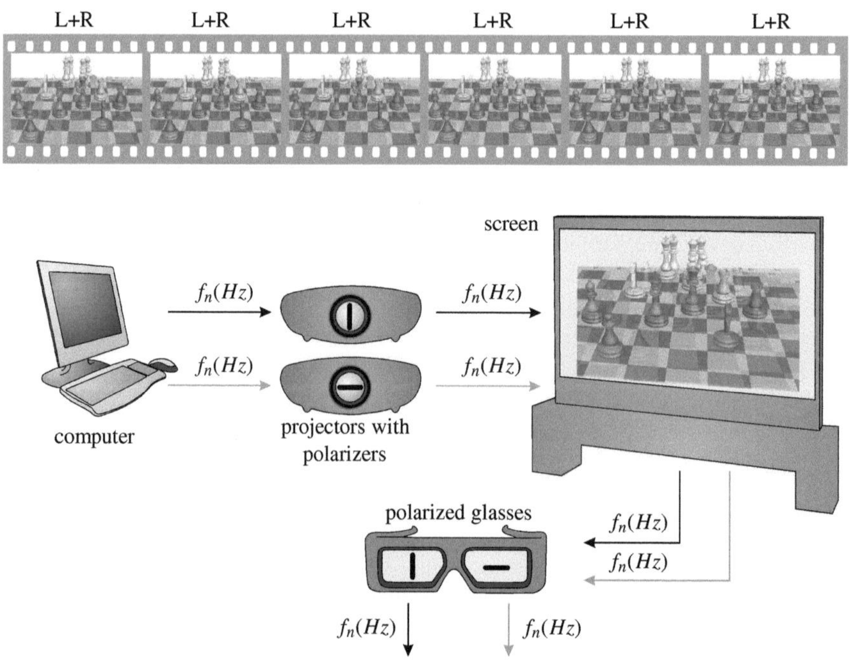

Fig. 5.26 Spatial multiplexing in a projection system

can be viewed by multiple people. Furthermore, polarized glasses are much lighter and cheaper than head-mounted displays. Examples of spatially multiplexed displays are shown in Figs. 5.25 (for an LCD) and 5.26 (for a projection system). In a projection system, two projectors are generally used, with each generating

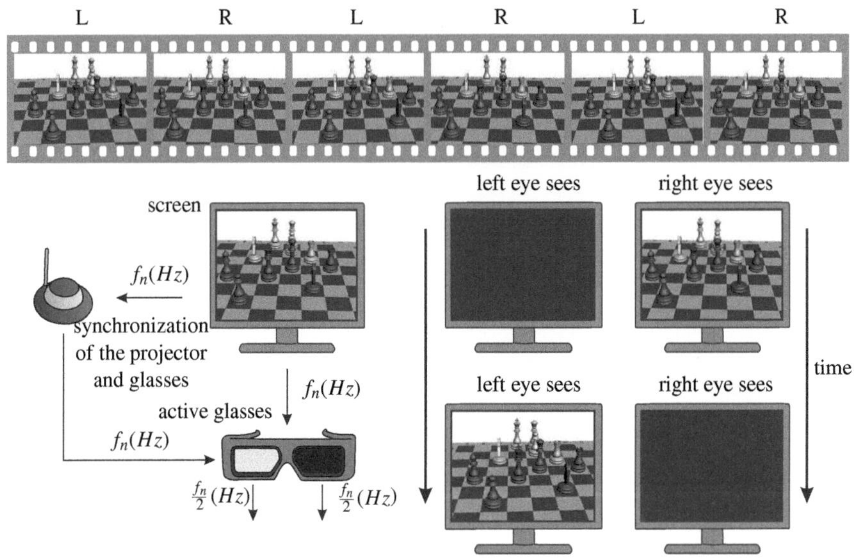

Fig. 5.27 Temporal multiplexing on an LCD screen

the part of the image for one eye. Two adjacent bands always show images for different eyes and are differently polarized. If the bands are sufficiently thin, a viewer with polarized glasses sees a single image with depth.

2. The image shown on the screen can also be temporally multiplexed. In this case, the screen rapidly switches between images for the left and right eyes. Only a single image is shown in a particular moment. If the switching frequency is very high, a viewer with active glasses perceives a single image with depth. The active glasses are synchronized with the display and block or allow the view of the screen depending on the currently displayed image. The best-known example of such polarized glasses is the 3D cinema where characters and object seem to 'jump' out of the image. Examples of temporal multiplexing are shown in Figs. 5.27 (LCD screen) and 5.28 (projection system).

5.3.3.4 Displays with Multiple Screens

The displays from the previous subsection provide a three-dimensional image, but it is limited to a single screen. The user thus cannot feel truly 'immersed' in a virtual environment since he/she is surrounded by a real environment from all other directions. However, better immersion can be achieved using multiple screens. The best-known example of such a system is the CAVE (Cave Automatic Virtual Environment)—a room in which the walls (and optionally the floor and ceiling) are covered with screens that display the virtual environment (Fig. 5.29). Each individual

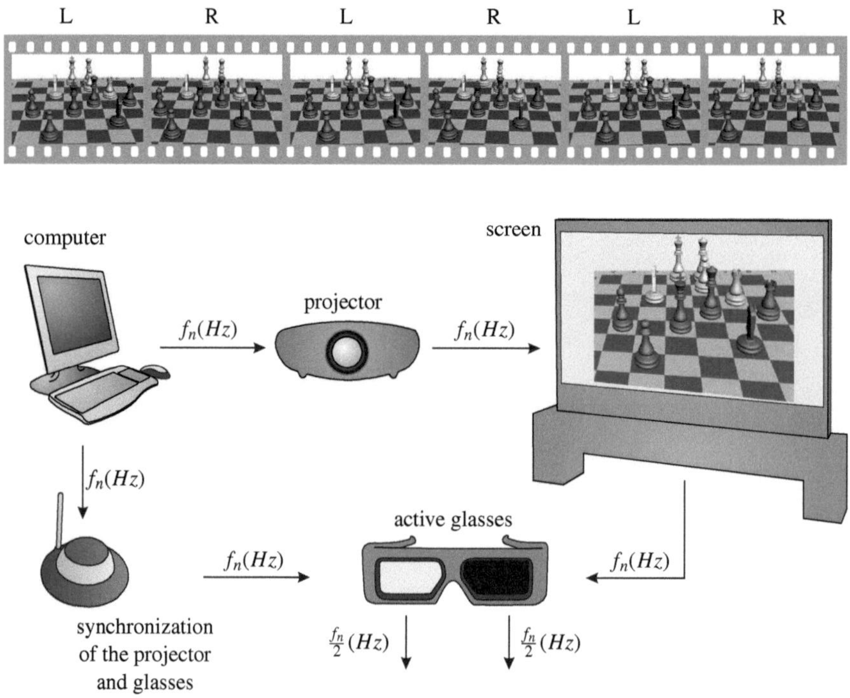

Fig. 5.28 Temporal multiplexing in a projection system

screen is a projection system just like those previously described, but different images of the environment must be generated for the multiple screens.

Systems such as the CAVE are almost always equipped with surround sound and motion trackers that adapt the images of the screens according to the user's position. They thus offer a strong feeling of presence in the virtual world. Their main disadvantages are size and price.

5.3.3.5 Autostereoscopic Displays

Autostereoscopic displays create an illusion of depth without the user needing to wear anything (e.g. glasses). Elements of the screen itself make the left eye see a different image than the right one. Today's displays mainly use two autostereoscopic technologies: parallax barrier and lenticular lenses. Both operate similarly to a display with polarized glasses and spatial multiplexing. The display simultaneously shows images for left and right eyes. In the case of the parallel barrier (Fig. 5.30), there is a barrier with a series of slits placed between the user and screen. The slits allow each eye to only see the parts of the screen that show the image meant for that eye. In the case of lenticular lenses, special lenses are placed in front of the screen. These

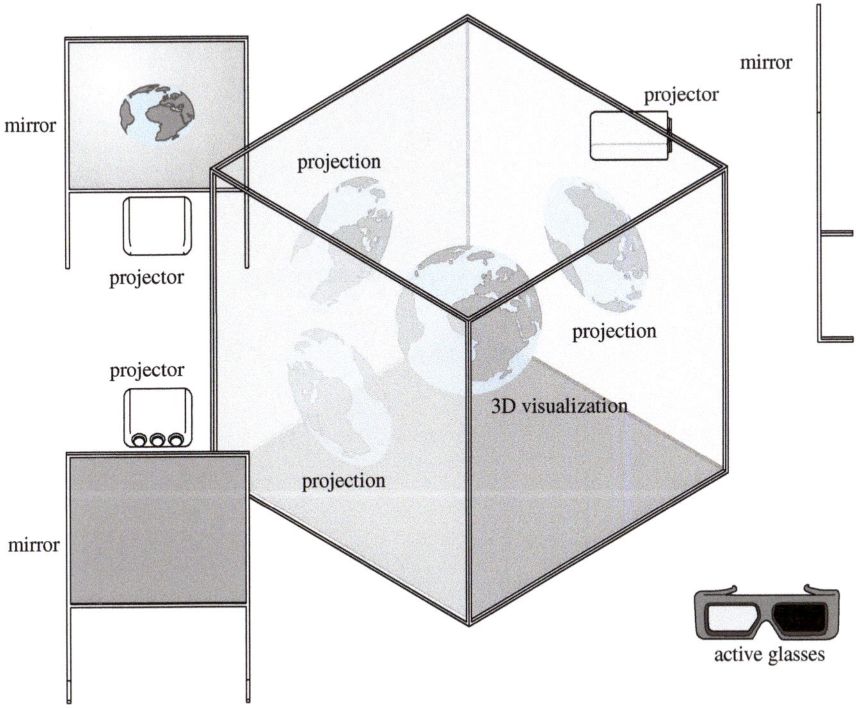

Fig. 5.29 Example of a CAVE system with three projection screens. The object in the *middle* of the CAVE system represents the virtual environment as perceived by the user

redirect light from the screen so that each eye only sees the light meant for that eye. In both cases, the main weakness is that the viewer needs to stand in a very specific location to correctly perceive depth. The weakness can be partially overcome using head tracking and corresponding image adaptation, but this increases the cost of the system and can only be done if there is only one viewer.

5.3.3.6 Volumetric Displays

All of the displays described thus far create an image on a screen in two dimensions. Volumetric displays go a step farther and create the image itself in three-dimensional space. The viewer does not only obtain an illusion of depth, but perceives real depth. The image can thus also be viewed from anywhere in the room. All volumetric displays are autostereoscopic, but deserve their own subsection due to their advantages. They are currently mostly used for military and research purposes, but can be expected to eventually achieve widespread use.

There are many types of volumetric displays. Perhaps the simplest method is a cube-shaped grid of small lamps that are transparent when turned off. Each lamp

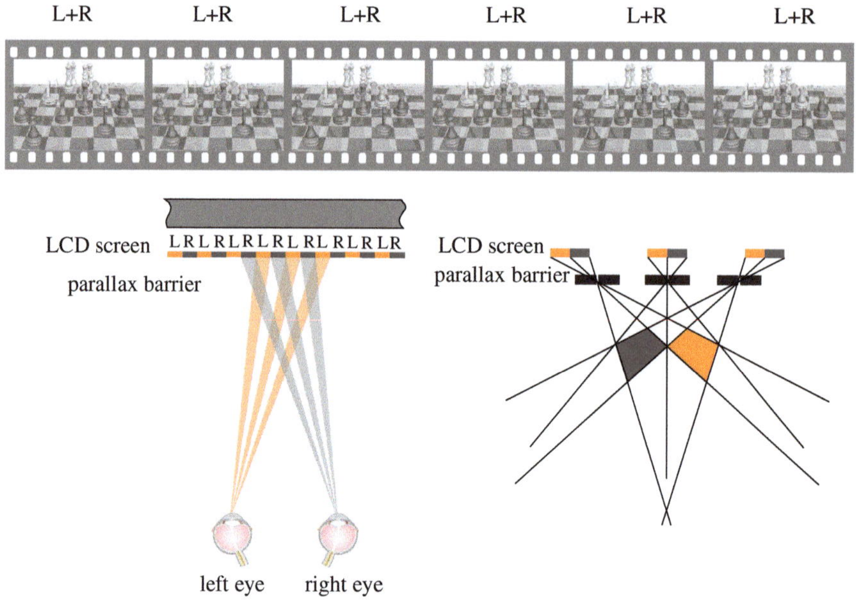

Fig. 5.30 Autostereoscopic display with a parallax barrier; the two polygons on the *right image* show the 'sweet spot' from which a three-dimensional image can be seen

represents a single voxel whose brightness and color can be individually tuned, thus building a three-dimensional image. This is also called the static method, as parts of the device do not physically move.

A second widespread method uses a moving base on which the image is displayed. An example is a rapidly rotating flat or spherical screen (Fig. 5.31). If the rotation is fast enough, a viewer perceives a three-dimensional image. The method is computationally demanding, as the three-dimensional virtual environment must very quickly be converted into consecutive two-dimensional images displayed on the screen.

The most famous volumetric method is holography: saving and reconstructing light rays reflected from objects. When the hologram is properly lit, it emits the same light as the original and thus represents a three-dimensional image that can be observed from any viewpoint. This method has been used to create images for decades, but did not appear in virtual reality. The reason for this delay is that dynamic images are much harder to display with a hologram than static ones. The principle of a hologram is as follows: light reflected from the object or scene we wish to record falls onto the surface of the hologram (Fig. 5.32). The hologram is simultaneously illuminated with a reference beam of light. The interference between both beams of light is recorded in the hologram. When the hologram is later illuminated with the same reference beam, diffraction of light on the recorded hologram results in the

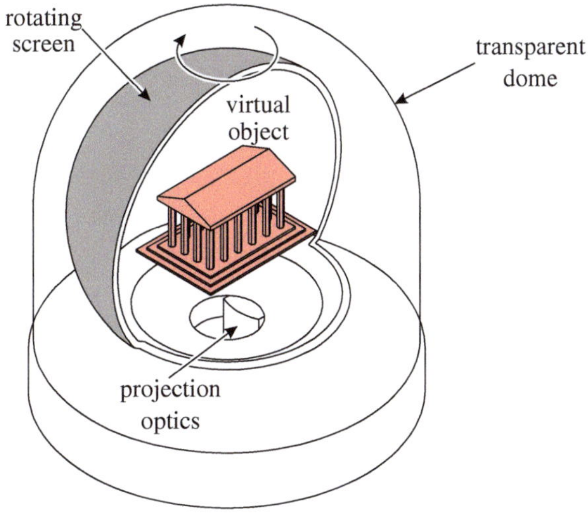

Fig. 5.31 Rotating screen and projection

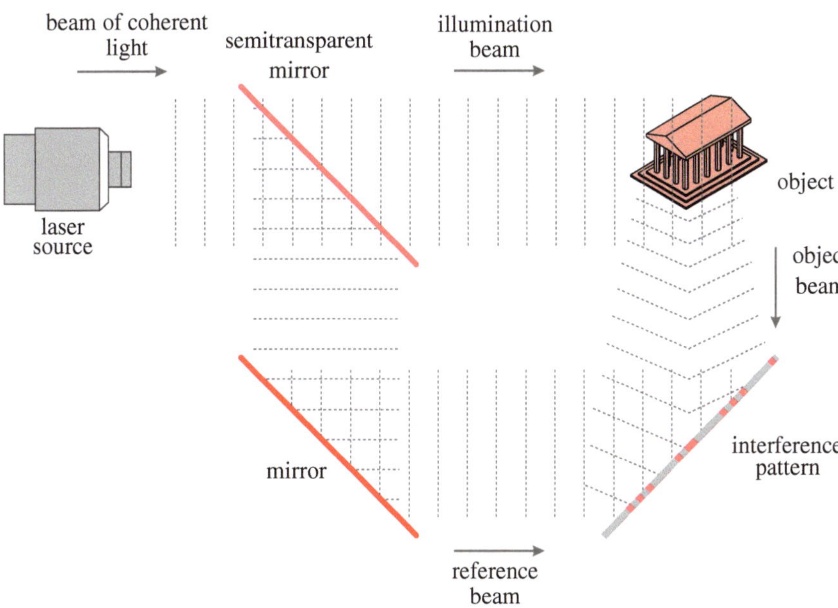

Fig. 5.32 The creation of a hologram

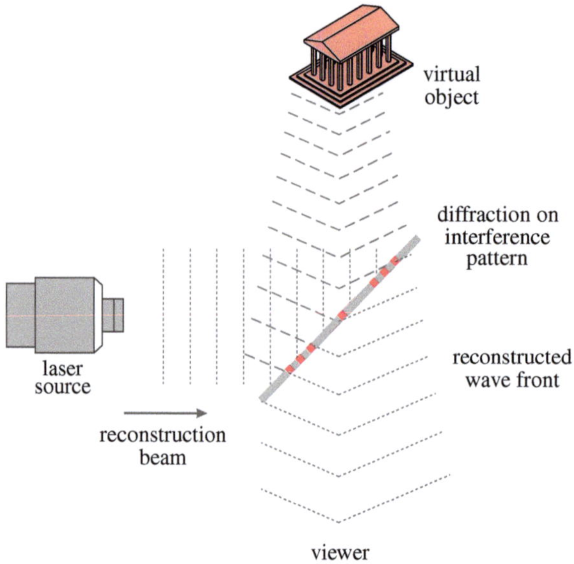

Fig. 5.33 The reconstruction of a hologram

same pattern of light that was reflected from the original image or scene (Fig. 5.33). The hologram itself is *not* an image; it is simply a pattern that can be illuminated in order to produce the recorded image.

Chapter 6
Acoustic Modality in Virtual Reality

Abstract Sound enhances the sense of realism in the virtual world, gives additional information about the environment, for example engine speed in flight simulators. By means of sonification the information is presented in the form of an abstract sound. Unlike vision, it is not limited to the direction of view, but is present regardless of head orientation. It is also not possible to temporarily disable your hearing the way it is possible to temporarily disable vision by closing your eyes. Temporal and spatial characteristics of sound differ from those of visual perception. Although what we see exists in space and time, vision stresses the spatial component of the environment. In contrast, hearing stresses the temporal component of the environment. In the following sections, the process of creating a virtual acoustic environment will be shortly presented with basic acoustic principles, human audio perception and recording techniques.

Sound enhances the sense of realism in the virtual world, gives additional information about the environment, for example engine speed in flight simulators. By means of *sonification* the information is presented in the form of an abstract sound. For example different temperatures of the building can be presented with different sound frequencies.

Sound not only attracts attention, but also helps determine the user's location. Like vision, hearing is a remote sense. Unlike vision, it is not limited to the direction of view, but is present regardless of head orientation. It is also not possible to temporarily disable your hearing the way it is possible to temporarily disable vision by closing your eyes.

Temporal and spatial characteristics of sound differ from those of visual perception. Although what we see exists in space and time, vision stresses the spatial component of the environment. In contrast, hearing stresses the temporal component of the environment. Since the main existence of sound is connected strictly to time, the speed of sound reproduction is more critical than the speed of image display. Sound can be divided into general ambient sound, sound that indicates events

M. Mihelj et al., *Virtual Reality Technology and Applications*,
Intelligent Systems, Control and Automation: Science and Engineering 68,
DOI: 10.1007/978-94-007-6910-6_6,

and sound that enhances or replaces other sensory perceptions. Ambient sounds are generally used to create mood in virtual reality. Sound can be presented as dematerialized background sound or sound from sound sources, i.e. entities in virtual space. The determination of the location of a sound source can effectively direct user attention. For example, loud sounds can cause a person to turn in the direction from which the sound is coming. Sound coming from a point in space can help create awareness about the existence of a certain object. This makes it possible for objects to create an illusion of still being present despite already being out of sight.

6.1 Acoustic Modality

In the following sections, the process of creating a virtual acoustic environment will be shortly presented with basic acoustic principles, human audio perception and recording techniques.

6.1.1 Process of Creating a Virtual Acoustic Environment

To create an acoustic virtual environment, also called a virtual acoustic display, three major design stages should be carried out [1] as shown in Fig. 6.1: definition, modeling and reproduction.

The *definition* of a virtual acoustic environment (VAE) includes foreknowledge about the sound sources, the room geometry and the listeners. *Auralization* is a term used to describe the rendering (modeling and reproduction) of audible sound fields [2]. Sound rendering is analogous to graphics rendering and is based on physical laws of sound propagation and reflection [3].

The *modeling* stage of sound rendering can be divided [1] into three tasks: source modeling, transmission medium modeling and receiver modeling.

Reproduction schemes used in implementing VAE can be divided [1] into three categories: binaural using headphones, binaural using loudspeakers with crosstalk cancelation, and multichannel reproduction.

6.1.1.1 Source Modeling

Source modeling is a concept used to produce sound in the VAE with properties such as directivity. The audio signals used in creation of the VAE include prerecorded digital audio samples and synthesized sounds. The audio source signals used should be 'dry', without any reverberant or directional properties. They are usually monophonic and regarded as a point source. If a sound source is a stereophonic signal, it is modeled as two point sources. The signal-to-noise ratio, sampling frequency and bit depth should be sufficient that the auralization does not produce undesired effects.

Sampling: A common way of generating sound is by playing a prerecorded sound originating from the physical world. The sampling frequency and number of sampling bits define the sampling resolution and the dynamic range of sound. Sampled sounds are especially suitable for generating realistic sound representation. They can be combined or changed, making it possible to produce a more abundant and less repetitive sound.

Sound synthesis based on computer algorithms allows greater flexibility in generating sound, but makes it harder to render realistic sounds. Sound synthesis is based on spectral methods, physical models or abstract synthesis. *Spectral methods* are based on sound spectrum observation and production of a sound that mimics the original. *Physical modeling* enables sound to be generated using models of physical phenomena, which result in a certain sound. The sound signal generated in this way is very realistic and can form an integral part of the environment. Sounds can simulate continuous or discrete events such as, for example, sounds of colliding objects. *Abstract sound* is usually original and not just a copy of a sound from the real world.

Postprocessing allows recorded sounds or sounds generated in real time to be additionally processed, which results in sounds similar to the original, but with certain qualitative differences. The added effects may be very simple, such as an echo (which illustrates that the sound comes from a large space) or attenuation of high frequency sounds (which results in an impression of the sound originating from a distance). *Echo* simulation further enriches the virtual world's sound characteristics. It is achieved by

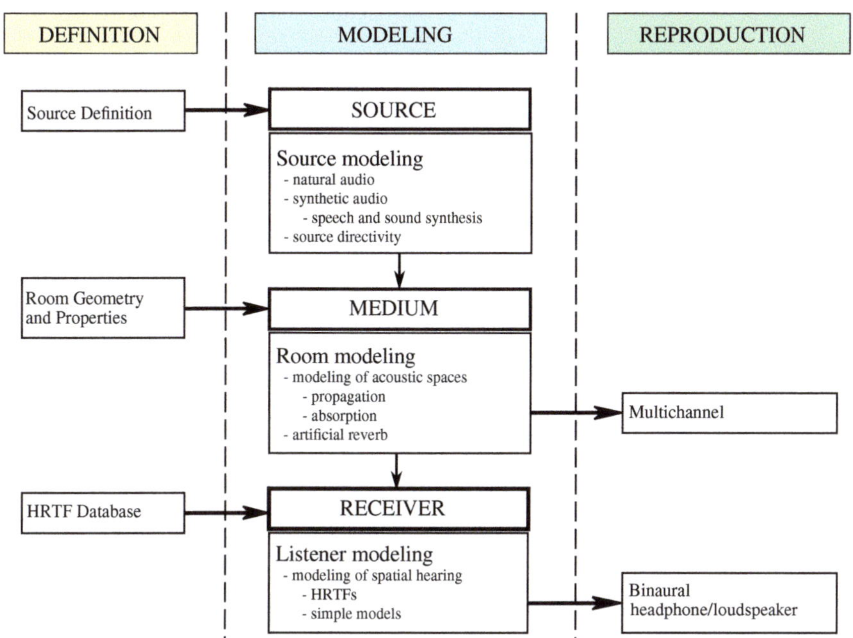

Fig. 6.1 Process of implementing a virtual acoustic environment, [1]

mixing the sound with a delayed and filtered sound that represents reflections from the walls of the room.

Sinusoidal sounds, described by mathematical sine functions, form the basis of spectral sound synthesis. Sinusoidal sounds are basic building blocks of computer-generated audio impressions, just as polygons are the basic building blocks of the visual domain. Since sounds are time-dependent, the equation that determines sound is also time-varying. The *frequency modulation* technique allows slightly richer sounds than spectral synthesis. In addition to the amplitude and frequency, sound contains additional parameters such as frequency of the carrier signal, the relationship between the modulation frequency and carrier signal frequency and modulation depth.

Additive and subtractive techniques of sound creation allow signals of different frequencies to be combined. The result therefore contains a rich combination of frequencies. The method essentially represents the addition of several sinusoidal signals of different frequencies with different phase shifts. The *Subtractive method* uses filters to attenuate certain frequency bands of an audio signal (e.g. white noise or harmonics). *Filtering* techniques allow the creation of effects such as demonstration of the place of sound origin or the acoustic characteristics of the room.

6.1.1.2 Transmission Medium Modeling

Modeling of room acoustics can be done by scale modeling or by computational simulation [1].

Scale modeling is usually used in the design of large concert halls and uses a three dimensional scaled down model of the hall and ultrasound sound sources.

In *computational modeling*, three different approaches can be used: ray-based modeling, wave-based modeling and statistical modeling.

- *Ray-based modeling* treats sounds like rays. This is possible when the wavelength of the sound is small compared to the area of the room. The roughness of the room surfaces should be small compared to the sound wavelength. Diffraction and interference phenomena are ignored. The simplified impulse response of the simulated virtual environment can be divided into three components: direct sound, early reflections and late reverberation. Ray tracing and the image-source method are the most commonly used methods and differ in how reflection paths are found.
- *Wave-based modeling* offers the most accurate result, but the analytical solution of the wave equation can be found only for some simple geometrical room shapes. Numerical wave-based methods should be applied in other cases.
- *Statistical modeling* methods are not suitable for auralization purposes, but mainly for the prediction of noise levels in coupled systems where sound transmission by structures is an important factor.

6.1.1.3 Receiver Modeling

The properties of human spatial hearing are modeled with the interaural amplitude difference, time delay and *head-related impulse response* (HRIR). More information about HRIR and *head-related transfer function* (HRTF) can be found in Sect. 6.4.1.

6.1.1.4 Reproduction Schemes

Some sound reproduction schemes are presented below [1, 4].

The *headphone-based* binaural reproduction scheme attempts to synthesize the appropriate sound field directly at the listener's ears. The headphone is the best possible reproduction system in this situation since it allows complete isolation between channels as well as good sound insulation from the surroundings. Spatial sound perception is based on the linear sound distortion due to its interaction with the head and torso. When listening with headphones, this interaction does not exist, so the information about the environment should be provided with a reproduced sound. This can be provided in two ways: by recording two signals with microphones located in the listener's ear (or with two microphones that simulate ears on an artificial head) or by processing a monophonic audio signal with the HRIR by convolution or in the frequency domain by multiplying the signal with the HRTF. Wearing headphones can be uncomfortable, especially for extended periods of time.

If sound signals are reproduced via headphones without any added spatial information, the sound will be perceived as being "inside the head". This is referred to as *inside-the-head-localization* [5]. The sound is perceived as moving inside the head with a tendency toward the backside of the head.

To ensure adequate sound pressure at the eardrum, the entire recording and reproduction chain needs to be calibrated. The effects of the headphones should be accounted for with *headphone equalization.*

Loudspeaker-based reproduction schemes:
Stereophonic reproduction is based on artificial positioning of sound sources between two loudspeakers, which are located at equal distances from the listener. The most important technique is the positioning of the sound by varying the amplitude of the playback *monophonic* recording through the loudspeakers. This technique can be extended with an additional third loudspeaker, which provides a virtual sound source placement above or below the horizontal acoustic plane, but still among the loudspeakers. A fully 3D sound field is created with more than three loudspeakers, which are placed on the surface of a virtual sphere surrounding the listener. The environment should not be too reflective. The virtual sound source in this configuration can only be positioned outside the virtual sphere, but can be in any direction between the three closest loudspeakers. Loudspeaker gain depends on the direction of the listener and the direction of the virtual sound source (which we want to simulate).

In *ambisonics*, the recording is done on four channels—W, X, Y, Z, with the sound image defined not only with intensity but also with the sound phase. W is a monophonic signal, X contains forward-backward information, Y contains left-right information and Z contains up-down information. If there is no Z channel, the overall sound picture is not destroyed. Only the height of the sound picture is truncated since the remaining signals W, X and Y will consistently transmit an authentic sound image in the horizontal plane with a full impression of the surrounding sound.

In the basic version, also known as *Ambisonic B format* the W channel represents a signal of a non-directional microphone while components X, Y, and Z are signals of microphones with a directional pattern in the form of eight directions. The W channel represents an absolute pressure while X, Y and Z represent pressure gradients in all three dimensions. The method does not depend on the reproduction system.

The goal of the *Wave Field Synthesis* method is to reproduce exact wave fronts inside the listening volume. The method is based on Huygens's principle, which says that any point of a wave front can be considered as a secondary source. If all these points are captured with a field of microphones, the entire wave front can be reconstructed with a field of loudspeakers. The drawback is the need for a large number of loudspeakers (and associated reproduction systems) while the advantage is that the perception of spatial audio image does not depend on the lister's position.

The *transaural* method uses a pair of loudspeakers for sound reproduction. In this case, a part of the signal intended exclusively for the left ear also reaches the right ear. Because of the *crosstalk*, the same thing happens with the sound from the right loudspeaker (Fig. 6.2). Our brain can not get the necessary information needed for 3D listening. In order to achieve 3D listening, an 'acoustic wall' should be built between the loudspeakers. This may be done with appropriate filtering, which introduces linear distortion. Filters allow 3D sound to be reproduced well at optimal points in space using only one pair of loudspeakers. It is necessary to ensure that the listener's head is in a fixed position since the area of crosstalk attenuation is limited. Otherwise, head position tracking is necessary in order to establish the position and orientation of the listener's head.

Only recently was a filtering method developed that allows 3D images to be reproduced without any additional sound coloration and with optimized crosstalk attenuation [6]. The term 'Pure stereo' describes a method of playback techniques that most closely approximates natural sound in the room. The technology was developed at Princeton University under the patronage of Professor Choueiri [6]. Its main advantage is the imitation of the original sound and imitation of its location in the room. Listeners need only one pair of loudspeakers to achieve full 3D sound reproduction. Unlike a regular stereo system, where the listening point lies midway between the loudspeakers, the Pure stereo system can freely move this point around the room using filter adjustments. The listener can still hear the recording outside the optimal point, but spatial perception is no longer as good as within or near the optimum point, the so-called *sweet spot*.

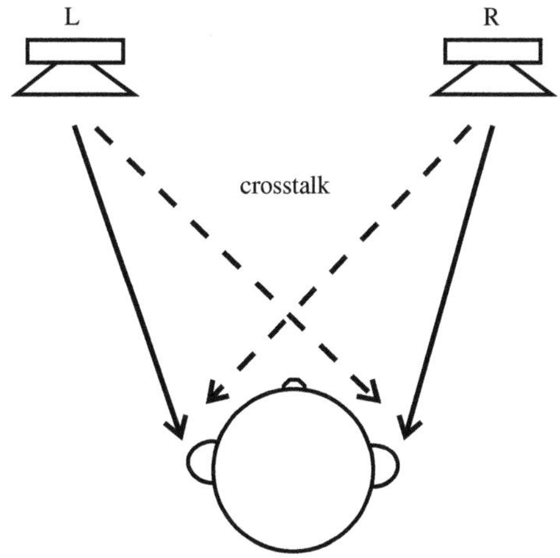

Fig. 6.2 Crosstalk in transaural reproduction

6.1.1.5 Computer Games and Virtual Reality

Computer games and virtual reality tend to make the gaming experience as realistic as possible. In addition to good graphics, a great emphasis is also placed on sound effects. Computer game developers are installing the latest technology for 3D audio reproduction in order to allow players to better immerse themselves into the game. The development of Pure stereo and other algorithms for stereo sound reproduction may provide a big step in the right direction for computer game and virtual reality fans.

6.1.2 Human Audio Localization

As virtual worlds are usually three-dimensional environments, it is important to have sound with three-dimensional features. *Localization* represents the psychoacoustic process of determining the distance and direction of the sound source either in the real or virtual world.

Auralization is used to create an illusion that the sound originates from a particular location in space. In order to achieve this, transfer functions can be used as a mathematical transformation tool that changes audio signals so that they appear to come from a specific point in space. Transfer functions are identified on the basis of measurements using a microphone in the listener's ear. The effect of filtering is set according to the direction of the sound source. Filter transfer functions determine the changes in the frequency spectrum of the signal when passing through the outer ear and various objects in the environment.

Sound localization is based on various characteristics. One of these is the ventriloquistic effect. *Ventriloquism* makes use of a psychoacoustics phenomenon resulting from the experience that the sound probably comes from the location from which it seems to have come. If someone hears a sound and sees the speaker's lips to be synchronized with the sound, (s)he expects the sound direction to be from the speaker's mouth. Ventriloquism is a strong characteristic that helps localize the sound.

In general, human localization of sound is a relatively underdeveloped ability. Therefore, strong and unambiguous localization characteristics should be used in virtual reality.

6.2 Fundamentals of Acoustics

Basic acoustic theory deals with vibrations and propagation of sound waves through different media and the effects caused by the propagating wave. Different areas of expertise explore different aspects of acoustics, making it a highly multidisciplinary science. Civil engineers are mainly interested in insulation and sound absorption in buildings. Architects are interested in room acoustics, which is mainly related to reverberation and echo studying in different halls. Electroacoustical engineers investigate the accuracy of sound transmission, conversion of electrical and mechanical energy into sound, and the design and construction of electroacoustic transducers. Physiologists examine the function and mechanism of hearing, auditory phenomena, and human reaction to sound or music. Psychoacoustics deals with human perception and interpretation of sounds. Linguists study the subjective perception of complex noises and cooperate with rehabilitation engineers in exploring the possibility of synthetic speech generation. Recently, more and more research has been conducted in the field of spatial sound with filter algorithms that enable listening to 3D spatial sound stereo systems.

To understand different aspects of acoustics and human hearing, we must first get acquainted with spatial sound from a physical and physiological perspective [7, 8]. Therefore, the next subsections present the basic characteristics of sound waves, the structure and functioning of the human hearing organ, as well as some basic concepts of room acoustics and its impact on the human sound perception.

6.2.1 Physics of Sound

Sound is a rapid pressure fluctuation with frequencies in the audible range of the human ear (approximately 20 Hz–20 kHz). Sound below this frequency range is called *infrasound* while sound above this range is called *ultrasound*. The sound is a mechanical wave that propagates through gases, liquids or solids that are at least a little compressible. They must therefore have the characteristics of persistency and elasticity. Persistency allows the fluctuation transfer from one particle to another particle while elasticity moves the particle to its steady state. Sound is always a longitudinal wave motion in gases, but can also appear as a transverse wave in liquids and solids. Any disturbance in a fluid medium is converted into fluid movement in the

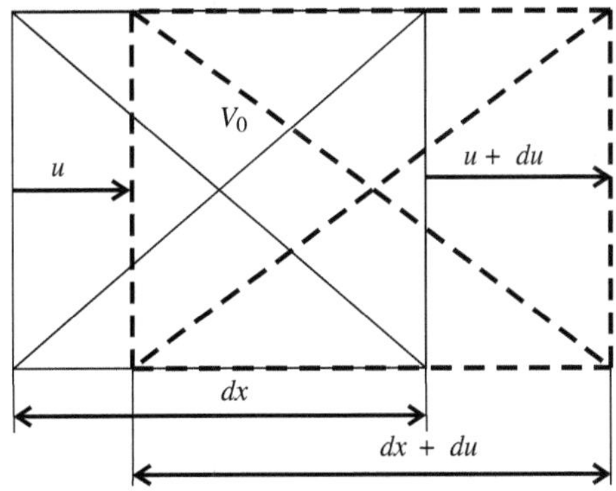

Fig. 6.3 Schematic representation of air volume changes in a tube

direction of wave propagation, producing small changes of pressure and density that fluctuate around the equilibrium state. When these compressions and rarefactions travel along the medium, they cause space- and time-dependent changes in pressure and density.

During compression, density and pressure are increased. During rarefaction, they are decreased. A relationship between pressure values and deformation can also be established in a segment of a medium. Equation (6.1) applies

$$m = \rho_0 \, V_0. \tag{6.1}$$

If the volume changes by dV, the density also changes by $d\rho$, Fig. 6.3. Considering the Eq. (6.1), we can write

$$m = \rho_0 \, V_0 = (\rho_0 + d\rho) \, (V_0 + dV). \tag{6.2}$$

The solution is in the form of

$$d\rho = -\rho \, \frac{dV}{V_0}. \tag{6.3}$$

Expression dV/V_0 defines the local relative change in volume. The sound field can be described by the velocity of the particle movements. We use Newton's law of $F = m \, a$, which can be written for fluids as

$$\frac{\delta p}{\delta x} = -\rho \frac{dv}{dt}. \tag{6.4}$$

This is actually the Euler equation for fluid physics in one-dimensional space. In general, it can be written

$$\nabla p = -\rho \frac{dv}{dt}. \tag{6.5}$$

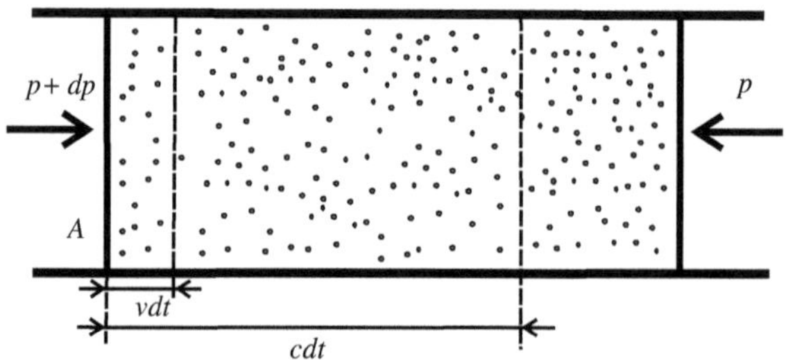

Fig. 6.4 Long cylindrical tube with cross-section A, density ρ and pressure p with piston on the left side

6.2.1.1 Calculation of Longitudinal Wave Propagation Speed

Pressure p on the left side of a pipe, Fig. 6.4, with cross-section A is increased with the help of a piston by dp. A compression therefore arises near the piston. The compression moves to the right with the speed of c. The average speed of the piston and the compression after the time dt is v. The momentum of air G has in the time dt increased by $dG = dm\, v = A\, c\, dt\, \rho\, v$ [9]. Impulse equals the momentum change.

The compressive impulse is $dG = F dt$, $dG = A c dt\, \rho v$, $dp = \rho c v$, $F = A dp$. The average speed reached by particles under pressure difference dp depends on the substance's compressibility. The initial volume of material was $V = A\, c\, dt$, but has shrunk during time dt by $dV = A\ v\ dt$. Compressibility of the material κ is, by definition,

$$\frac{dV}{V} = -\kappa\, dp. \tag{6.6}$$

Followed by $\kappa\, dp = \frac{A\, v\, dt}{A\, c\, dt} = \frac{v}{c}$ and $v = \kappa\, c\, dp = \rho\, \kappa\, c^2\, v$. We have obtained a final equation for longitudinal wave propagation speed

$$c = \sqrt{\frac{1}{\kappa\, \rho}}. \tag{6.7}$$

In the air the *speed of sound* (sound wave propagating through the medium), is given with the equation

$$c = \sqrt{\frac{\vartheta p}{\rho}}, \tag{6.8}$$

where ϑ is the specific heat of the air ratio at constant pressure c_p with the specific heat at constant volume c_v, p is the pressure in N/m^2, ρ is the density in kg/m^3. At room temperature of 20 °C and normal atmospheric pressure of 1,000 mbar and

60 % relative humidity, the speed of sound is 344 m/s. At different temperatures, the speed of sound can be analytically calculated using the equation

$$c = c_{20}\sqrt{1 + \frac{T[^{\circ}\mathrm{C}] - 20}{273}}. \tag{6.9}$$

For a quick approximate calculation we can assume that the speed increases by 0.6 m/s per each degree Celsius.

6.2.1.2 Wave Equation

The wave equation describes the sound pressure change as a function of place and time. For one dimension, it has the form

$$\frac{\partial^2 p}{\partial x^2} - \frac{1}{c^2}\frac{\partial^2 p}{\partial t^2} = 0, \tag{6.10}$$

where p is the acoustic pressure and c is the speed of sound in the air. If speed c is constant, then the general solution of the Eq. (6.10) is equal to

$$p = f(ct - x) + g(ct + x), \tag{6.11}$$

where f and g are functions that can be derived twice. For a sine wave traveling in one dimension, we take one of the functions f or g sine function and the other is zero. We get a solution in the form

$$p = p_0 \sin(\omega t \mp kx), \tag{6.12}$$

where p_0 is the sound pressure amplitude, ω is the angular frequency of waves, k is the wave number $k = 2\pi/\lambda$ and λ is the wavelength.

6.2.1.3 Audio Intensity

The sound source performs work by exciting the oscillation of particles in the matter. The particle oscillation spreads through the substance in the form of sound energy. Sound wave energy is composed of the kinetic and potential energy of the particles oscillating under the influence of waves. For a sinusoidal wave, the wave energy per volume unit of material (energy density) through which the wave spreads is

$$\frac{\rho v_0^2}{2} = \frac{\rho(y_0\omega)^2}{2} = \frac{\rho\omega^2 y_0^2}{2}; \tag{6.13}$$

where y_0 is the amplitude of particle oscillation and ω is the angular frequency of the oscillation. We are also interested in wave energy flux density, which is the amount of energy transported by wave motion in a unit of time through one unit of cross-sectional area. If a sinusoidal wave travels at a speed c through the medium, it is

$$I = \frac{\rho c \omega^2 y_0^2}{2}. \tag{6.14}$$

Sound wave energy flux density is proportional to the square of the amplitude and frequency of the oscillating particles. Sound intensity I is preferably given as amplitude of the pressure difference Δp rather than offset amplitude y_0. If the relationship between the pressure and the amplitude is considered, $dp = \rho\, \omega\, y_0\, c$, $(dp)^2 = \rho^2\, c^2\, \omega^2\, y_0^2$, $I = \frac{dp^2}{\rho\, c\, 2}$, $p_{ef} = \frac{dp}{\sqrt{2}}$ we get

$$I = \frac{p_{ef}^2}{\rho c}, \tag{6.15}$$

where p_{ef} is root mean square of pressure level in N/m^2, ρ is the density in kg/m^3, c is the speed of sound in the m/s. At room temperature and normal atmospheric pressure is $p_{\text{ef}} = 2 \cdot 10^{-5}$ N/m^2 and the density is $\rho = 1.21$ kg/m^3. The speed of sound is 343 m/s and sound intensity is about 10^{-12} W/m^2.

6.2.1.4 The Density of the Sound Energy

The density of the sound energy is the energy per one volume unit in the observed substance. The potential energy of sound waves comes from the substance displacements and the kinetic energy comes from the particle movement. If there are no losses, the sum of the two energies is constant. The current density of the sound energy is

$$E_{tr} = \rho \dot{x}^2 + \frac{p_0 \dot{x}}{c}. \tag{6.16}$$

The average density of the sound energy is

$$E_{pop} = \frac{1}{2} \rho \dot{x}^2. \tag{6.17}$$

where ρ is the current volume weight in kg/m^3, p_0 is the static pressure in N/m^2, $\dot{x}$ is the speed of the particle in m/s, and c is the speed of sound in m/s.

6.2.1.5 The Specific Acoustic Impedance

The specific acoustic impedance of matter is defined as the ratio of sound pressure and particle velocity

$$z = \frac{p}{v}, \tag{6.18}$$

where p is sound pressure in N/m^2 and v is the speed of the particle in m/s. In a standing wave, the acoustic impedance varies from point to point. In general, it is a complex number

$$z = r + ix, \tag{6.19}$$

where (r) is specific acoustic resistance and x is specific acoustic reactance.

6.2.2 Audio Measurements

A logarithmic scale is used for sound power, intensity and pressure measurements because of the large measurement range. *The sound power level* (*PWL, Power Level*) is defined as

$$PWL = 10 \log \frac{W}{W_0}, \tag{6.20}$$

where W is power in W and W_0 is the reference power in W. The standard reference for power is $W_0 = 10^{-12}$ W. *The Sound Intensity (IL)* is defined similarly

$$IL = 10 \log \frac{I}{I_0}. \tag{6.21}$$

The standard reference sound intensity is $I_0 = 10^{-12}$ W/m^2. The *Sound Pressure Level* (*SPL*) is defined as

$$SPL = 20 \log \frac{p}{p_0}. \tag{6.22}$$

The standard sound pressure reference is $p_0 = 2 \cdot 10^{-5}$ N/m^2.

Intensity Spectrum Level (ISL) and *Pressure Spectrum Level (PSL)* are defined as

$$ISL = 10 \log(\frac{I}{I_0} \Delta f), \tag{6.23}$$

$$PSL = SPL - 10 \log(\Delta f). \tag{6.24}$$

Spectrum intensity level is the noise level at some frequency, defined as the level of noise intensity in a given frequency band width of 1 Hz and with a centre frequency of f. *PSL* is defined similarly to *SPL*, within a frequency band of 1 Hz.

6.2.2.1 Acoustic Power Measurements

We roughly distinguish three methods for measuring acoustic power in air: free field measurements, diffuse field measurements and the comparative method.

Total power emitted by the source in an open, *free field* is obtained by integrating the sound intensity all over the area that surrounds the origin. In practice, a sphere is used as the measuring surface and is closed with a reflective bottom. The sound source is in the centre of the hemisphere, and the radius should be at least twice the largest dimension of the source. The problem is the choice of measuring locations, especially if the source emits pure tones or if it is oriented. In that case, a large number of measuring points should be selected.

Sound excited in a closed reflective space causes the occurrence of *diffuse sound fields*. In this case, the specific measurement surface can no longer be defined. The power emitted by the source in the oscillating state equals the power absorbed by the walls of the room.

Acoustic power is determined by *comparing* the measured sound source to a reference sound source with a known power level or acoustic power level. The emitted signal must be broadband and spectrally evenly distributed and should not have strong directional characteristics. The measurement procedure can be carried out in a closed or open area.

Acoustic power determination through the intensity measurement: The method is based on the fact that there is a link between the intensity of the sound I and cross power density $K_{12}(\omega)$ between two sound pressure values, measured by two microphones, placed at some distance apart.

In addition to this method, another method is also used where the volume of the sound pressure phase gradient in the sound field is determined. Practical implementation is possible using a single microphone. In this manner, problems due to phase differences of the two-channel version can be avoided.

6.2.3 Room Acoustics

Records concerning echo and sound absorption can be found in the manuscripts from middle ages. The founder of modern acoustics, Wallace Clement Sabine (1868–1919), was a physicist who began acoustic development at Harvard University in order to improve classroom acoustics. With the help of some organ flutes (used as sound sources), a stopwatch and his skilled hearing, he conducted the first scientific measurements in the history of acoustics.

Various factors have an impact on room acoustics or on the desired sound transmission from the acoustic source to the listener: the room volume, the room shape and the distribution of different absorbing materials. Important room acoustic parameters are the reverberation time RT_{60}, Eq. (6.25), and the succession of the first acoustic reflections.

In an open space, we usually deal only with the direct sound field. The basic characteristic of open space is that sound pressure drops by 6 dB when the distance is doubled (law $1/r$). In confined spaces, the impact of the diffuse sound field caused by reflections and scattering from the walls is also important.

The *reverberation time* RT_{60} is defined as the time in which the original sound energy falls to the millionth part of its value (a drop of 60 dB). Since 1900, different equations for RT_{60} calculation have been proposed. The most prominent is still the Sabine's equation

$$RT_{60} = \frac{4\ln(10^6)}{c}\frac{V}{Sa} \doteq 0.161\frac{V}{Sa} \tag{6.25}$$

and

$$a = \sum_{i=1}^{n} \frac{S_i a_i}{S}, \tag{6.26}$$

where V is the volume of space in m^3, c is the speed of sound in the air, S is the total surface area in m^2, S_i is the specific surface area in m^2, a_i is the absorption coefficient of each surface, and a is a mean absorption coefficient. The absorption coefficient a_i is, by definition, the proportion of absorbed acoustic power with respect to the incident acoustic power.

6.3 Sound Perception

Man's hearing is optimized for listening in a three-dimensional sound field. It consists of two physically separate sensor systems: the left and the right ear. The detection mechanism using two ears is called *binaural listening*. In order to understand the complexity of listening, the anatomy and physiology of the ear will be introduced first.

6.3.1 Anatomy and Physiology of the Ear

The human ear is divided into three functional parts: the external, middle and inner ear. Figure 6.5 shows the cross section and allocation of the ear.

The outer ear consists of the pinna and the external auditory canal. The pinna consists of skin and cartilage and serves to guide the waves in the external auditory

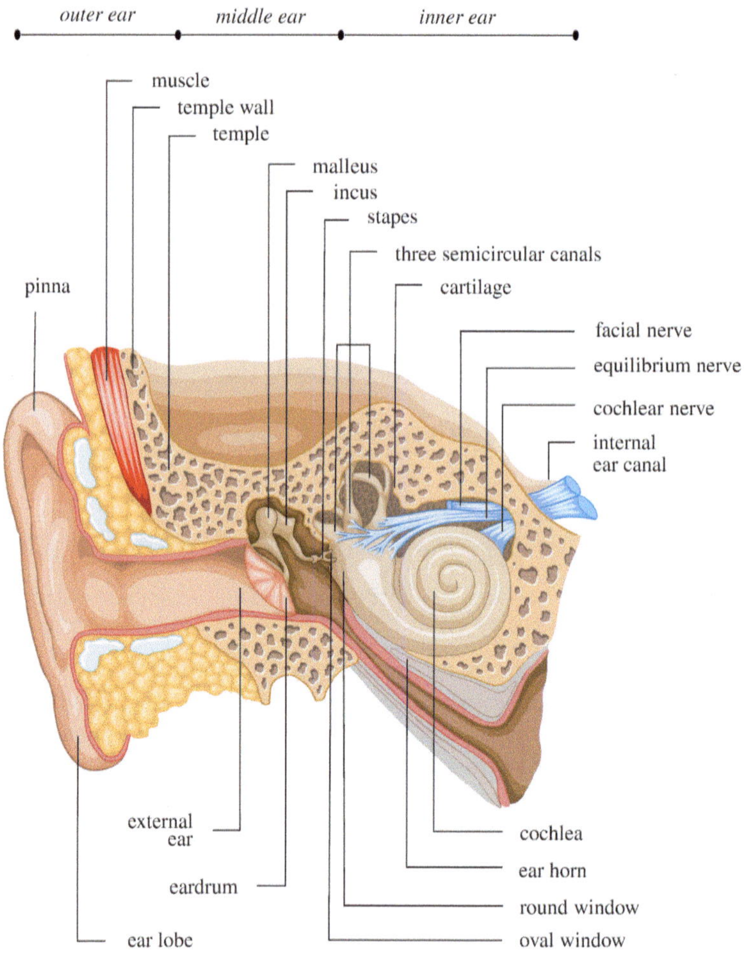

Fig. 6.5 Anatomy of the human ear

canal. Th external auditory canal amplifies frequencies in the range from 2 up to 5 kHz. It is about 3.5 cm long and has a diameter of 0.5–1 cm. At the end of the ear canal there is the eardrum.

The eardrum, Fig. 6.5, separates the external auditory canal and the *middle ear*, which is connected with the outside world via the Eustachian tube. The eardrum and oval window are linked by three ear bones: the hammer (malleus), anvil (incus) and stirrup (stapes). Their task is to transmit the eardrum vibrations to the inner ear and to provide for adjustment of amplitude and force of eardrum and oval window oscillation. The muscles attached to the middle ear bones are involved in the regulation of the transmitted vibration volume.

The inner ear consists of two sensory organs: a body balance organ and a cochlea. The cochlea is a tube wrapped two and a half times around a bone pillar. The cochlea cavity is divided into two parallel channels: the upper (or atrial) channel and the lower (or eardrum) channel. The atrial channel is connected to the atrium and through the oval window to the middle ear. The eardrum channel is adjacent to the middle ear through the round window. The cochlea is made up of three sections, two of which are filled with perilymph. In the middle of them is the endolymph-filled cavity. The channels are separated from each other by thin membranes (Reissner and basilar). The organ of Corti lies on the basilar membrane and consists of sensory hair cells, neurons and several types of supporting cells. It is named after its discoverer, Alfonso Corti (1822–1876). Sensory hair cells are connected with the nerve fibers of the auditory nerve.

Active mechanics: Operation of a living cochlea, as opposed to a dead one, depends on the active mechanical process with a positive feedback loop that amplifies the basilar membrane response. The amplification is carried out by the outer sensory hair cells. Most of the information (90–95 %) comes from the cochlea to the brain via internal sensory hair cells, although any damage of outer hair cells leads to hearing damage.

6.3.2 Loudness

Loudness is the extent of auditory sensations caused by sound reaching the human ear. Vibrational energy is a physical property while loudness is based on psychological interpretation. Loudness is therefore a subjective quantity and as such can not be accurately measured. Usually a relative measure based on a logarithmic ratio of two amplitudes is used. Man can hear sounds only in a certain range of frequencies and sound pressure levels. The human hearing range is from 20 Hz to 20 kHz, but the upper limit decreases with aging.

Figure 6.6 shows equal-loudness contours that are perceived by man as equally loud. Equal-loudness contours were first published in 1933 by Fletcher and Munson [10] and later by Churcher and King [11], but these measurements show small discrepancies. In 1956 they were corrected by Robinson and Dadson using a new contour for the lower auditory threshold [12]. Free field equal-loudness contours were standardized by ISO 226:1987, Fig. 6.6, and later revised in 2003. The ratio of sensitivity to high and low tones depends on the intensity of the sound waves. Maximum human ear sensitivity is in the range of 2–5 kHz. The external auditory canal has a resonance frequency at approx. 3 kHz. Subjective sound impression depends on the frequency content or spectrum of the sound and on the sound amplitude.

As the human ear perceives sound waves of the same power as differently loud if different frequencies are in the sound wave, it is necessary to introduce the loudness level with the unit *phon*.

Phon is an acoustic measure used to indicate the general noise loudness level. Pure tone with a frequency of 1000 Hz at a sound intensity level $IL = 1$ dB has, by definition, the volume level of one phon. All other tones have a sound level of n phons if the ear judges them to be as loud as a pure tone of frequency 1000 Hz with the sound intensity level of n dB. A tone with a frequency of 500 Hz and a volume level of 40 phons sounds as loud as any other tone of 40 phons with a freely chosen frequency.

Sone is an acoustic criterion used in determining sound loudness level. It is used to compare and classify the loudness of different sounds based on the way the ear hears them. By definition, a pure tone with a frequency of 1000 Hz has a sound loudness of one son at a sound intensity level of 40 dB. A loudness of 1 milison represents the hearing threshold.

The *Loudness Level (LL)* is defined by Eq. (6.27) as

$$LL = 10 \log \frac{I}{10^{-12}}, \tag{6.27}$$

where I is the intensity of sound in the $\mathrm{W/m^2}$. The link between loudness in sons and loudness in phons can be written by the equation

$$S = 2^{\frac{P-40}{10}}, \tag{6.28}$$

where S is loudness in sones and P is loudness in phons.

Noy was proposed in acoustics to assess and compare the level of noise or annoyance noise as perceived by the ear. The ear subjectively judges whether a certain noise is louder, but not necessarily more annoying than another, not so loud noise.

A human ear in the normal human hearing frequency range perceives sound when the amplitude of the pressure difference is at least 10^{-4} μbar ($10^{-5}\mathrm{N/m^2}$) and a

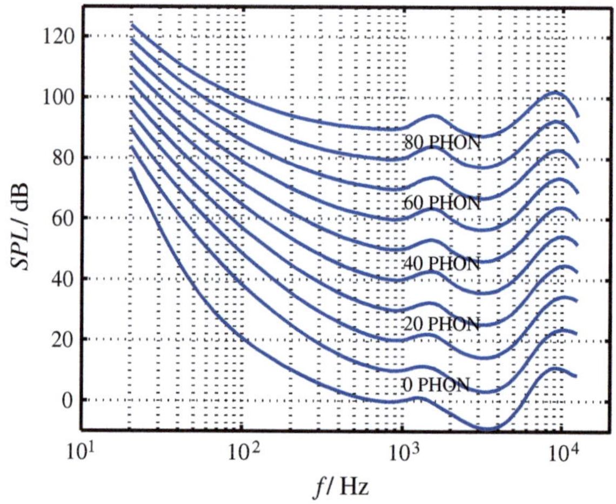

Fig. 6.6 Equal-loudness contours according to standard ISO 226:2003

max. of 1 mbar (the eardrum survives this pressure difference without pain). The energy that must reach the ear at a certain frequency for the tone to still be heard was measured after World War I. The lower auditory threshold (minimum sound intensity) to excite sensation at 3 kHz was established around 10^{-12} W/m^2. On the other hand, the density of sound that the ear can still handle without pain is around 1 W/m^2. The human ear is therefore a very sensitive instrument, detecting sound waves in the range of 12 orders os sound intensity. However, the sensitivity of the human ear is not the same across the whole range. Experiments show that sensitivity is proportional to the logarithm of sound intensity, Fig. 6.7.

The sensitivity of the ear for sound power level $L_0 = 10^{-12}$ W/m^2 is zero because the sound is too weak to be perceptible to the ear. Sound level between the lower threshold and the threshold of pain is divided into 130 phones.

6.4 The Spatial Characteristics of Hearing

To define the spatial dimensions of sound perception, the listener's head is placed in a coordinate system as shown in Fig. 6.8. Angle φ represents the azimuth and δ represents the angle in the vertical plane (elevation).

The ability of the human ear to identify and localize the direction of a sound source with great precision is called *auditory localization* or *binaural audition*. Auditory localization is conditioned by the difference of sound intensity in both ears and is caused by diffractions, reflections and the phase difference in the sound that comes at different times to both ears.

A person is able to determine the direction of a sound source, except in the case of a plane wave in a free field coming from the direction where $\varphi = 0°$. When the sound source is right in front of us, sound waves reach the left and right ear simultaneously. The direction then cannot be accurately determined. As soon as the head is moved out of the symmetry plane, the sound wave arrives to one ear earlier and the direction can be obtained from the timing difference.

Our hearing apparatus uses a number of sound properties in order to determine the origins of sound. These properties are the result of the propagation of sound waves from the sound source to the listener's eardrum. The head, being a natural barrier in the sound field, causes reflections and refractions of sound waves. With its geometry, it has an additional impact on the sound field at higher frequencies

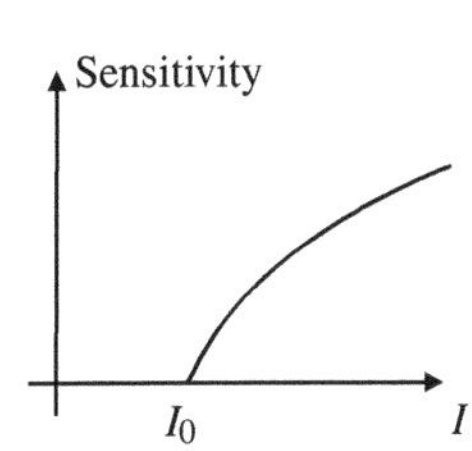

Fig. 6.7 Sensitivity of human hearing depending on sound intensity level

in relation to its dimensions and its shape compared to the sound wavelength size. The influence of the room on the sound wave is expressed by absorption, reflection, refraction and interference due to items located on the path between the sound source and the listener etc. The influences of the surroundings are important because they give information about the distance from the sound source to the listener. Reflections, diffractions, refractions and interference of sound waves due to the listener's body (especially shoulders, head and ears) have a key impact on determining the sound source direction (azimuth and elevation). All this causes a difference between the sound pressure coming to the left and sound pressure coming to the right ear. These differences also depend on the direction from which the sound comes to the human head (φ and δ).

6.4.1 Binaural Hearing

Human ears are about 18 cm apart, allowing sound direction to be determined and enabling stereo hearing. When the sound source is outside the listener's frontal plane time and intensity differences occur between the ears as the sound to one ear travels around the head [13]. The model to calculate the time difference is shown in Fig. 6.9. The time difference between ears Δt is

$$\Delta t = \frac{r\varphi + r\ \sin(\varphi)}{c} = \frac{r(\varphi + \sin(\varphi))}{c}, \tag{6.29}$$

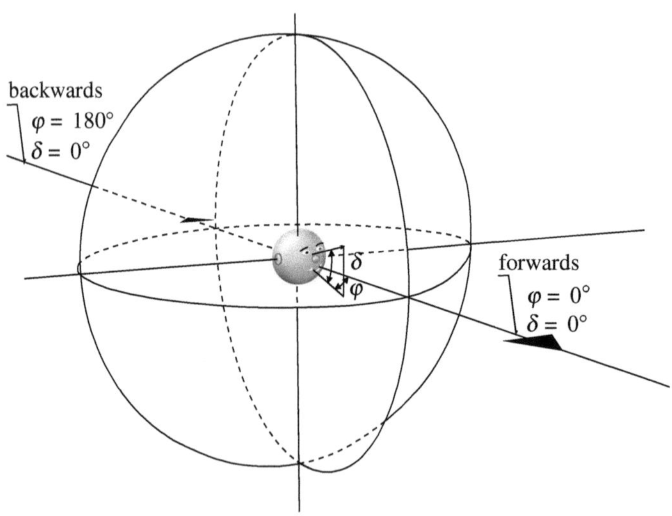

Fig. 6.8 The head placed in the sound perception coordinate system

where r is the head radius, φ is the deviation of the sound source from the frontal plane and c is the speed of sound. The maximum time difference occurs at an angle of $\varphi = 90°$ or $\pi/2$ radians. Taking into account Eq. (6.29), it is equal to

$$\Delta t_{\max} = \frac{r(\varphi + \sin(\varphi))}{c} = \frac{0.09\,\mathrm{m}(\frac{\pi}{2} + \sin(\frac{\pi}{2}))}{344\,\frac{\mathrm{m}}{\mathrm{s}}} = 673\,\mu\mathrm{s}. \tag{6.30}$$

This small time difference, which varies with the angle φ, defines the sound wave phase difference and thereby the frequency by which the sound source direction can be determined. The phase difference ϕ can be calculated as

$$\phi = 2\pi f \,\frac{r(\varphi + \sin(\varphi))}{c}. \tag{6.31}$$

When the phase difference becomes greater than 180° or π radians, the sound direction becomes indistinguishable with this method since at that angle there are two possible sound source positions on the left and on the right side. Considering this limit, we can calculate the maximum frequency at a certain angle φ, that is

$$f_{\max}(\varphi) = \frac{c}{2 \cdot 0.09\ \mathrm{m} \cdot (\varphi + \sin(\varphi))}, \tag{6.32}$$

which at an angle $\varphi = \pi/2$ gives us

$$f_{\max}(\varphi = \pi/2) = \frac{344\ \mathrm{m/s}}{2 \cdot 0.09\ \mathrm{m} \cdot (\pi/2 + \sin(\pi/2))} = 743\,\mathrm{Hz}. \tag{6.33}$$

This means that at an angle of $\varphi = 90°$, the highest frequency for which we can still determine the direction of the sound origin is 743 Hz. For smaller angles φ the maximum frequency increases [13].

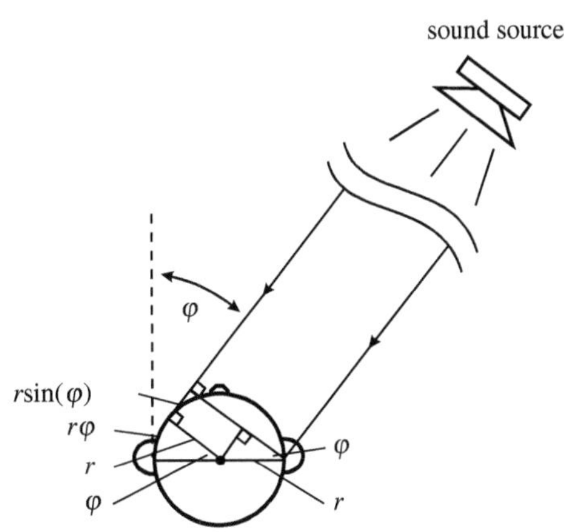

Fig. 6.9 Model for time differences caused by the sound outside the frontal plane, calculation [13]

The second method of sound direction perception originates in the sound intensity difference caused by the shading effect. When the sound source is outside the frontal plane, the amplitude of the sound is reduced (shaded) at the ear that is further away from the source. Experiments have shown that the amplitude ratio between the two ears varies sinusoidally between 0 and 20 dB, depending on the angle of the sound source and the frequency or wavelength. When the wavelength is longer than the object (head), the sound wave refracts and diffracts around the object. The wavelength therefore has virtually no impact on the wave spreading. At wavelengths smaller than the object, there is almost no refraction and the object has a huge impact on the wave propagation. The size of the object at which the wavelength becomes an important factor in sound wave propagation is about two thirds of the sound wavelength (2/3 λ), although the scattering of sound waves already starts at frequencies an octave below. That means that the minimum frequency at which we are able to detect the direction of the sound source is when the diameter of the head is about one third of the wavelength (1/3 λ). Taking into account the width of the head $d = 18$ cm and the angle at which the opposite ear is most shadowed ($\varphi = \pi / 2$), the minimum frequency equals

$$f_{\min\ (\varphi=\pi/2)} = \frac{1}{3}\frac{c}{d} = \frac{1}{3}\frac{344\,\mathrm{m/s}}{0.18\,\mathrm{m}} = 637\,\mathrm{Hz}. \tag{6.34}$$

It follows thereby that the direction perceived from the difference in sound intensity is useful for higher frequencies and the direction perceived from timing differences is useful for low frequencies. The above method of determining the sound source position is impossible if the sound source is at the same angle φ placed in front or behind the listener, because the calculation from Eq. (6.29) shows that the time difference Δt is the same in both cases.

The complex form of the pinna causes delays in the received audio signal, which are a function of sound source direction in all three dimensions. The delays are so small that this effect comes into force only at frequencies greater than 5 kHz. Another equally important way of determining the direction of the sound source is by turning the head. In this way, the sound signals coming from the back move in a different direction compared to the audio signals coming from the front [13].

All of the above methods of sound direction detection create an impression of three-dimensional sound in the listener's head—binaural listening.

The listener's ability to determine the direction of the incoming sound is based on the difference of the sound pressure that comes to the left and to the right ear. When the sound comes to our ears from one direction, time and intensity differences of sound pressure levels are created. Time differences in the sound pressure level between the ears are called interaural time differences. The dependence of interaural delay on the azimuth angle φ for an average sized head is shown in Fig. 6.10. This of course applies to the sound wavelengths, which are large in relation to the head's dimensions (the distance between the ears). The direction of the sound source is determined on the basis of the phase delay between the ears (depending on the wavelength). This time difference is large enough for our brains to determine the direction from which the sound is coming. The human ear can detect the time difference of 30 μs, which corresponds to the change of the sound source azimuth angle $\varphi = 3°$.

For phase delays larger than 180° of phase angle or π radians, the direction becomes indistinguishable as there are two possible sound source positions: on the left or on the right side of the head. This occurs at a frequency of about 1 kHz.

Let's look at the situations at 40 Hz and at 1,000 Hz when the distance between the ears is $L = 0.18$ m. The sound is coming from a direction in such a way that: $\varphi = 90°$ and $\delta = 0°$

$$f_1 = 40\ \text{Hz}$$
$$\lambda = \tfrac{c}{f} = \tfrac{340\ \text{m/s}}{40\ \text{Hz}} = 8.5\ \text{m}$$

$$\varphi = \frac{L}{\lambda} = \frac{0.18\ \text{m}}{8.5\ \text{m}}\,360° = 7.6°$$

$$f_2 = 1\ \text{kHz}$$
$$\lambda = \tfrac{c}{f} = \tfrac{340\ \text{m/s}}{1000\ \text{Hz}} = 0.34\ \text{m}$$

$$\varphi = \frac{L}{\lambda} = \frac{0.18\ \text{m}}{0.34\ \text{m}}\,360° = 190.6°$$

At low frequencies (below 100 Hz), the phase differences are so small that the sound source direction cannot be determined. This means that the sound pressure levels on both ears are almost identical. Our brain thus can not determine the direction of the sound (Fig. 6.11). This property is exploited in multi-channel loudspeaker systems, where only one subwoofer is required. Similarly, when compressing stereo audio data, information from the low-frequency spectrum can be combined into one channel.

The head acts as a barrier at frequencies of the same or much larger wavelength than the wavelength of sound. Therefore, differences in the sound pressure level occur between the ears.

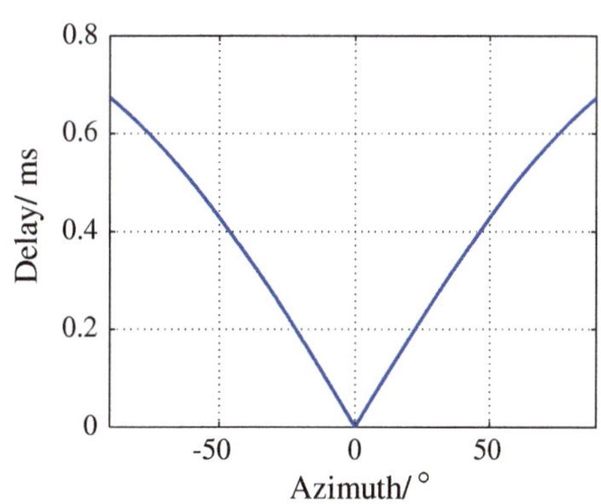

Fig. 6.10 Dependence of interaural delay on the azimuth angle φ for an average-sized head

Human perception of sound direction is based on a linear sound distortion due to the interaction of sound with the listener's head and torso. The influence of the head, body and pinna on the sound propagation is described in the collection of responses of the transmission path between the sound source and the listener's eardrum on unit impulse. This response is called HRIR (Head Related Impulse Response). Its Fourier transform is HRTF (HRTF—Head Related Transfer Function). HRIR contains all the necessary information to determine the direction of the sound source—Interaural Time Difference—ITD, Interaural Level Difference—ILD and amplitude characteristics. The transfer functions depend on the azimuth and elevation angles. Figure 6.12 shows the HRTF and HRIR for a test person in the position where $\varphi = 90°$ [14]. We should be aware that each person has her/his own HRTF to which the person has adapted her/his hearing.

Interaural delay time and amplitude interaural differences are of crucial importance for sound source localization in the horizontal dimension (azimuth) while differences between HRTFs at different elevations are crucial for localization in the vertical direction (elevation). Variations in HRTFs become evident at frequencies above 4 kHz, when the wavelength of sound waves becomes comparable to the size of the pinna. The pinna affects the spectral properties of sound primarily with the so-called *pinna notches*, which depend on the elevation of the perceived sound and occur between 5 and 17 kHz. This is consistent with the argument that, for the precise elevation localization, the perceived sound should contain high-frequency components. The localization accuracy in the middle and horizontal plane is shown in Fig. 6.13 [15].

Humans are nevertheless capable of sound localization when the sound contains only frequencies of up to 3 kHz. The presence of higher frequencies improves the localization accuracy, but the effects of reflections from the body, which appear below 3 kHz, are sufficient for successful elevation localization. The frequency content of the sound has an impact on azimuth localization as well. It also turns out that a wider spectrum of sound provides a more precise localization. Human hearing phenomena

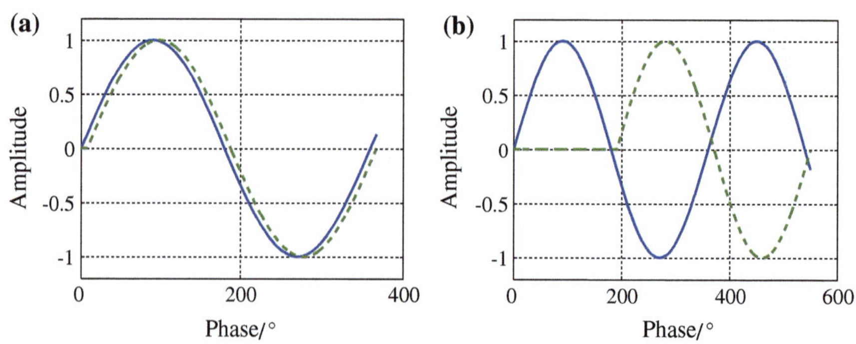

Fig. 6.11 The phase delay between the ears at a frequency of 40 Hz (**a**) and 1 kHz (**b**)

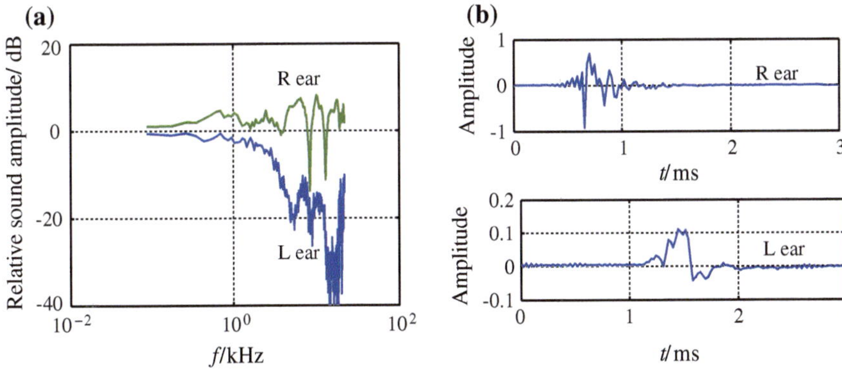

Fig. 6.12 HRTF (a) in HRIR (b) at azimuth $\varphi = 90°$

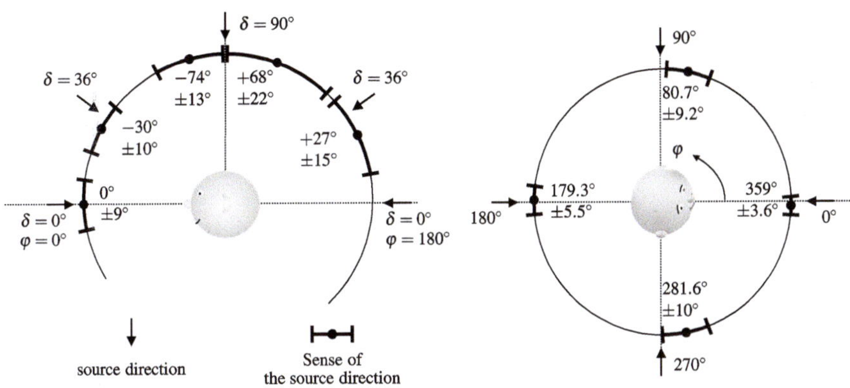

Fig. 6.13 The localization accuracy in the *middle plane* (*left*) and *horizontal plane* (*right*) [15]

in relation to the elevation and distance have been relatively poorly explored for the time being, while the azimuth has been far better studied [16].

The human auditory system has limited ability to determine the distance of a sound source, see Fig. 6.14 [15]. To assess the distance from the sound source, human hearing relies on the property that high-frequency sound is more strongly attenuated in the air than low-frequency sound. Distant sounds thus have more pronounced basses than high tones. We hear remote sources with lower strength than the surrounding sound sources. If the sound source or the listener is moving, the Doppler effect arises. The relationship between the direct and the reflected sound gives us information about the distance to the sound source. The impression of distance disappears when tones in the sound field last for a long time.

In a closed space, the ratio between the direct and the reflected sound assists us in determination of the sound source direction. As a result of binaural listening, human hearing has the ability to distinguish direct sound from reflected sound and automatically gives more weight to the direct sound.

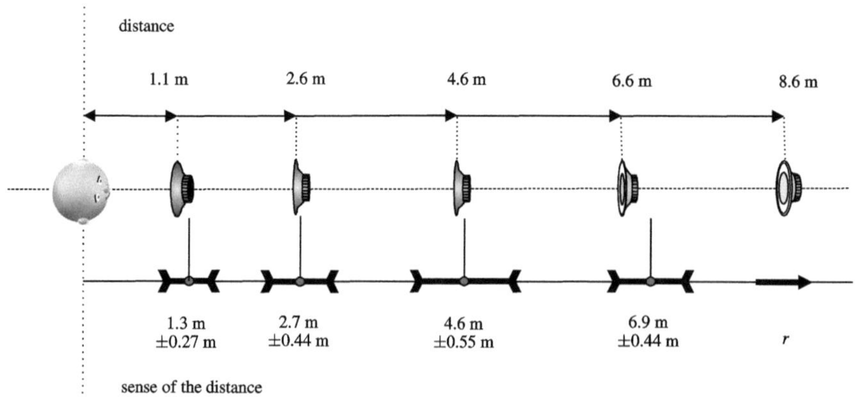

Fig. 6.14 The sense of the sound source distance for sound pulse in the horizontal plane and azimuth 0° [15]

In a complex sound image, *auditory masking* occurs. It can occur in the frequency domain (simultaneous, frequency or spectral masking) or in the time domain (temporal masking or non-simultaneous masking).

Human hearing range is limited by the lower auditory threshold and the upper pain limit. Human hearing cannot distinguish amplitude levels lower than 1 phon. Detailed measurements have shown that audio sources in the sound field which are more than 15 phons weaker than the loudest sound source can be ignored. However, if two equally loud sound sources are present in the sound field, the overall volume level of the two is likely to increase over the theoretical value of 3 dB up to 10 dB. Due to nonlinearity of human hearing, a new scale has also been introduced. Sound loudness expressed in *sones* is distinguished by the feature that it corresponds to n times greater impression of loudness. At low loudness, sones increase more slowly than phones. When two distinct tones are present in the sound field and the stronger does not cover the weaker one, the total volume can be approximated by the sum of the sones.

As soon as a tone is heard, the sensitivity for weaker tones at different frequencies is lowered. Therefore the hearing threshold rises, see Fig. 6.15. For frequencies above 1200 Hz, the masked tones should be amplified by 40–50 dB to be audible again.

If the tone mix (music) loudness is increased, lower frequency tones are heard better.

Time perception of sound: Audio tones of the same power, if heard for a different amount of time, are not perceived as equally loud. Measurements have shown that the loudest sound impression is made by a tone if listened to for a time span of 0.5 to 1.5 s.

The overlap effect: When we speak, sound waves are conducted to our own ear both via bones and via air. These two channels overlap, which is why we do not recognize our own voice recorded on a tape, even though others find it a faithful reproduction.

A tone is perceived as a continuous phenomenon. The ear should receive at least 3 whole sound periods. Otherwise, the sound is heard only as rapidly following claps.

Human ear distortion: The complex ear system has different sensitivities to different frequencies and therefore introduces *linear distortion* to the audio signals. The relationship between the tone phases also changes. The human auditory system also introduces *nonlinear distortion*, which is why the difference or sum of tones arriving to the ear can be heard. Since low-frequency tones can mask higher-frequency tones, high-frequency tones can in turn mask the low-frequency tones only if they are substantially amplified.

Audio *beat* interference occurs when two audio tones with equal amplitude and slightly different frequencies are present. It is perceived as a periodic volume variation.

Binaural listening allows us to focus on one sound source in the presence of other sounds. This phenomenon was first studied by E. Collin Cerry in 1953 and named the *Cocktail Party Effect* [17]. Even better results are achieved if the sound sources are distributed in space around us. For example, you can easily understand and focus on one speaker in a multitude of other equally loud speakers.

Each signal can be extracted from the mixture (or mixtures) of signals on the basis of statistical properties using Independent Component Analysis. The properties of the signals do not necessarily need to be known (blind signal separation), the only restriction is that they are statistically independent and no more than one signal can be noise [18].

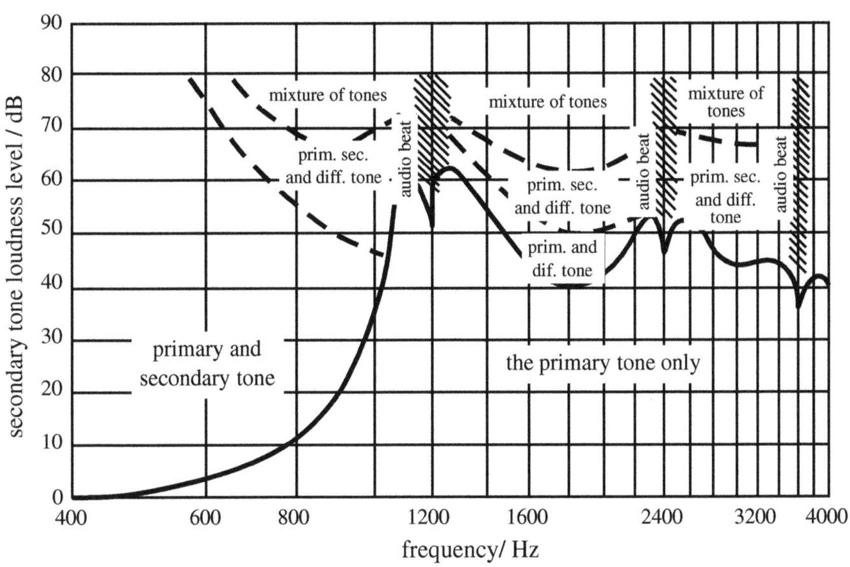

Fig. 6.15 Auditory masking with the primary tone at the frequency of 1200 Hz and amplitude of 80 dB

The amazing brain flexibility and its ability to replace missing senses is reflected also in the fact that humans, like some animals (bats, dolphins, ...), develop echo-localization. Echo-localization is a way of navigation or observation by means of sound using reverberation. Sound is reflected from objects and in turn creates a picture of the environment in the brain according to the reflected sound.

6.5 Recording Techniques

In addition to good-quality reproduction, stereo effect allows us to get an impression of space—artists' spatial layout (in playing orchestra for example).

6.5.1 Recording in Stereophony

The main recording techniques used in stereophony are the XY and MS techniques [19]. A special microphone combination *coincident microphone* is used for recording. In practice, two identical condenser microphones are located close to each other, and their directional characteristics can be changed by changing their polarization voltage. The horizontal plane can be adjusted by the upper microphone rotation. An intensity difference between the two output channels occurs when both directional characteristics of this system are different or when the microphones are oriented in different directions. Three basic types of microphone characteristics are used: circular, cardioid and figure-of-eight.

Coincident microphones can be used for MS and XY techniques. In the MS technique, the major axis of the M microphone points towards the center of the sound source while the S microphone points in parallel direction to the source (S). In the XY system, the microphone axes are placed at 90° to each other and 45° to the sound source. The conversion between both techniques can be carried out by simple addition and subtraction:

$$M + S = X \tag{6.35}$$

$$M - S = Y \tag{6.36}$$

$$X + Y = M \tag{6.37}$$

$$X - Y = S \tag{6.38}$$

The current stereo technique is compatible, which means that the monophonic reception is as acoustically accurate as optimal monophonic recording.

6.5.1.1 Binaural Technique of 3D Sound Recording

A dummy human head (mannequin head) with microphones built in the ear canals is used for binaural recording. It must be reproduced with headphones to create 3D stereo sensation for the listener and give the impression of actually being in the acoustic scene. Unlike 'Pure stereo' playback technique, the binaural technique does not require any additional filtering. If played back with loudspeakers, the 3D spatial information is corrupted due to the crosstalk.

References

1. Savioja L, Huopaniemi J, Lokki T, Vaananen R (September 1999) Creating interactive virtual acoustic environments. J Audio Eng Soc 47(9):675–705
2. Kleiner M, Dalenback BI, Svensson P (1993) J Audio Eng Soc 41(11):861–875
3. Lokki T, Savioja L, Väänänen R, Huopaniemi J, Takala T (2002) Creating interactive virtual auditory environments. IEEE Comput Graph Appl 22(4):49–57
4. Andre C, Embrechts JJ, Verly JG (2010) Adding 3d sound to 3d cinema: identification and evaluation of different reproduction techniques. In: International conference on audio language and image processing (ICALIP), pp 130–137
5. Kapralos B, Jenkin MR, Milios E (2008) Virtual audio systems. Presence Teleoper Virtual Environ 17(6):527–549
6. Choueiri E (2010) Optimal crosstalk cancellation for binaural audio with two loudspeakers.Posted on the 3D3A Lab's website
7. Seto WW (1971) Theory and applications of acoustics. Mcgraw-Hill, New York
8. Jeglič A, Fefer D (1991) Osnove akustike. Akademska založba Ljubljana
9. Kladnik R (1966) Fizika za slušatelje tehniških fakultet, vol 1. Fakulteta za arhitekturo, gradbeništvo in geodezijo
10. Fletcher H, Munson WA (1933) Loudness, its definition, measurement and calculation. J Acoust Soc Am 5(2):82–108
11. Churcher BG, King AJ (1937) The performance of noise meters in terms of the primary standard. J Inst Electr Eng 81:57–90
12. Robinson DW, Dadson RS (1956) A re-determination of the equal-loudness relations for pure tones. Br J Appl Phys 7:166–181
13. Howard DM, Angus J (2001) Acoustics and Psychoacoustics, 2nd edn. Oxford
14. Listen HRTF database: http://recherche.ircam.fr/equipes/salles/listen/
15. Ivančević B (2007) Elektroakustika, 3rd edn. Sveučilište u Zagrebu Fakultet elektrotehnike i računarstva. ISBN 978-953-184-118-4
16. Sušnik R (2006) Postopki kodiranja elevacije izvorov zvoka v akustični sliki prostora. Ph.D. thesis, Fakulteta za elektrotehniko, Ljubljana
17. Beguš S (2001) Slepo ločevanje akustičnih izvorov, diplomsko delo
18. Lee TW (1998) Independent component analysis: theory and applications. Kluwer Academic Publishers, Boston
19. Fefer D, Jeglič A (1992) Elektroakustika. Fakulteta za elektrotehniko in računalništvo

Chapter 7
Haptic Modality in Virtual Reality

Abstract The chapter covers topics relevant for the design of haptic interfaces and their use in virtual reality applications. It provides knowledge required for understanding complex force feedback approaches and introduces general issues that must be considered for designing efficient and safe haptic interfaces. Human haptics, mathematical models of virtual environment, collision detection, force rendering and control of haptic devices are the main theoretical topics covered in this chapter, which concludes with a summary of different haptic display technologies.

The word *haptic* originates from the Greek verb *hapto—to touch*—and therefore refers to the ability to touch and manipulate objects. The haptic experience is based on tactile senses, which provide awareness of the stimuli on the surface of the body, and kinesthetic senses, which provide information about body pose and movement. Its bidirectional nature is the most prominent feature of haptic interaction, which enables exchange of (mechanical) energy—and therefore information—between the body and the outside world.

The word *display* usually emphasizes the unidirectional nature of transfer of information. Nevertheless, in relation to haptic interaction, similar to visual and audio displays, the phrase *haptic display* refers to a mechanical device for transfer of kinesthetic or tactile stimuli to the user.

Virtual environments that engage only the user's visual and auditory senses are limited in their ability to interact with the user. It is desirable to also include a haptic system that not only transmits sensations of contact and properties of objects, but also allows their manipulation. The human arm and hand allow objects to be pushed, grasped, squeezed or hit, they enable exploration of object properties such as surface texture, shape and compliance, and they enable manipulation of tools such as a pen or a hammer. The ability to touch, feel and manipulate objects in a virtual environment, augmented with visual and auditory perception, enables a degree of immersion that otherwise would not have been possible. The inability to touch and feel objects, either

M. Mihelj et al., *Virtual Reality Technology and Applications*,
Intelligent Systems, Control and Automation: Science and Engineering 68,
DOI: 10.1007/978-94-007-6910-6_7,

in a real or a virtual environment, impoverishes and significantly affects the human ability of interaction with the environment [1].

A haptic interface is a device that enables interaction with virtual or physically remote environments [2, 3]. It is used for tasks that are usually performed by hand in the real world, such as manipulating objects and exploring their properties. In general, a haptic interface receives motor commands from the user and displays the appropriate haptic image back to the user.

Although haptic devices are typically designed for interaction with the hand, there are a number of alternative options that are appropriate for sensory and motor properties of other parts of the body. In general, a haptic interface is a device that: (1) measures position or contact force (and/or their time derivatives and spatial distribution) and (2) displays contact force or position (and/or their spatial and time distribution) to the user.

Figure 7.1 shows a block diagram of a typical haptic system. A human operator is included in the haptic loop through a haptic interface. The operator interacts with a haptic interface either through force or movement. The interface measures human activity. The measured value serves as a reference input either to a teleoperation system or a virtual environment. A teleoperation system is a system in which a (usually) remote slave robot accomplishes tasks in the real environment that the human operator specifies using the haptic interface. Interaction with a virtual environment is similar, except that both the slave system and the objects manipulated by it are part of the programmed virtual environment. Irrespective of whether the environment is real or virtual, control of the slave device is based on a closed-loop system that compares the output of the haptic interface to the measured performance of the slave system. The essence of haptic interaction is the display of forces or movements, which are the result of the operation of the slave system, back to the user through the haptic interface. Therefore, it is necessary to measure forces and movements that occur in teleoperation or compute forces and movements that are the result of interaction with a virtual environment. Since force may be a result of movement dynamics or interactions of an object with other objects or with the slave system, collision detection represents a significant part of the haptic loop. As already mentioned, contact can occur either between objects in the environment (real or virtual) or between an object and the slave system. Collision detection in a real environment is relatively straightforward and is essentially not much more than the measurement of interaction forces between the robot and its surroundings. In contrast, collision detection in a virtual environment is a more complex task since it requires computation of contact between virtual objects that can be modeled using different methods. In this case, it is necessary to compute multiple contacts between external surfaces of objects.

Collision detection forms the basis for computation of reaction forces. In a teleoperation system, force is measured directly using a force/torque sensor mounted on the slave robot end-effector. In a virtual environment, on the other hand, it is necessary to compute the contact force based on a physical model of the object. The object stiffness can, for example, be modeled as a spring-damper system while friction can be modeled as a force that is tangential to the surface of the object and proportional

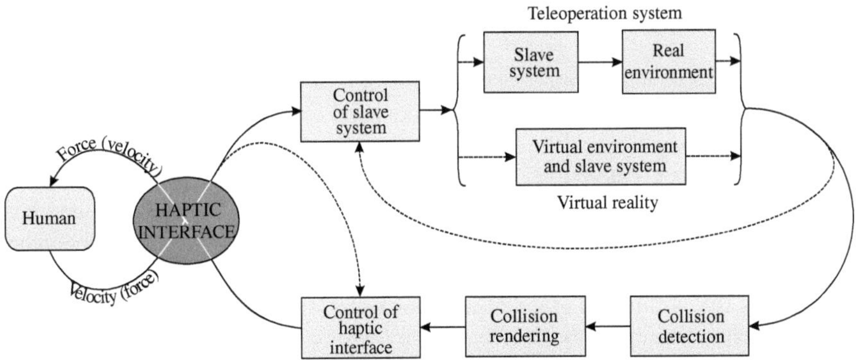

Fig. 7.1 Haptic system: interaction between a human and the haptic interface represents a bidirectional exchange of information—a human operator controls the movement of a slave system as well as receives information about the forces and movements of the slave system through the haptic interface

to the normal force to the surface of the object. The computed or measured force or displacement is then transmitted to the user through the haptic interface. A local feedback loop controls the movement of the haptic interface so that it corresponds to the measured or computed value.

From the block scheme in Fig. 7.1, it is clear that the interaction between a human and the haptic interface represents a bidirectional exchange of information—a human operator controls the movement of a slave system as well as receives information about the forces and movements of the slave system through the haptic interface. The product of force and displacement represents mechanical work accomplished during the haptic interaction. Bidirectional transfer of information is the most characteristic feature of haptic interfaces compared to display of audio and visual images. The following sections provide a general course on haptics in virtual reality. More information can be found in [4].

7.1 Human Perceptions and Motor System

Haptic perception represents active exploration and the process of recognizing objects through touch. It relies on the forces experienced during touch. Haptic perception involves a combination of somatosensory perception of patterns on the skin surface and kinesthetic perception of limb movement, position and force. People can rapidly and accurately identify three-dimensional objects by touch. They do so through the use of exploratory procedures, such as moving the fingers over the outer surface of the object or holding the entire object in the hand. The concept of haptic perception is related to the concept of extended physiological proprioception according to which, when using a tool, perceptual experience is transparently transferred to the end of the tool.

When designing haptic interfaces with the aim of providing optimal interaction with the human user, it is necessary to understand the *roles* of the motor, sensory and cognitive subsystems of the human haptic system. In the real world, whenever we touch an object, external forces act on the skin. Haptic sensory information, conveyed from hands to the brain during contact with the object, can be divided into two classes: (1) *Tactile information* refers to the perception of the nature of contact with the object and is mediated by low-threshold mechanoreceptors in the skin (e.g. in the finger tip) within and around the contact area. It enables estimation of spatial and temporal variations of the distribution of forces within the area of contact. Fine texture, small objects, softness, slipperiness of the surface and temperature are all perceived by tactile sensors. (2) *Kinesthetic information* refers to the perception of position and movement of a limb together with the forces acting on that limb. This perception is mediated by the sensory nerve signals from the receptors in the skin around the joints, in the joints, tendons and muscles. This information is further augmented with motor control signals.

Whenever arm movement is employed for environment exploration, kinesthetic information enables perception of natural object properties such as shape and compliance or stiffness. Information transmitted during passive and stationary hand contact with an object is predominantly tactile (kinesthetic information provides details about the position of the arm). On the other hand, during active arm motion in free space (the skin of the hand or of the arm is not in contact with surrounding objects) only kinesthetic information is conveyed (absence of tactile information indicates that the arm is moving freely). In general, both types of feedback information are simultaneously present during actively performed manipulation tasks.

When actively performing the task, supervision of contact conditions is as important as the perception of touch. Such supervision involves both fast muscle or spinal reflexes and relatively slow voluntary responses. In motor tasks such as pinch grasp, motor activity for increasing grasp force occurs in as little as 70 ms after the object slip is detected by fingertips. Human grasping and manipulation skills are the result of mechanical properties of skin and subcutaneous tissue, rich sensory information from diverse and numerous receptors that monitor the execution of tasks and the ability of nervous system to fuse this information with the activity of the motor system.

In addition to tactile and kinesthetic sensory subsystems, the human haptic system is composed of a motor system, which enables active exploration of the environment and manipulation of objects, and a cognitive system, which associates action with perception. In general, contact perception is composed of both tactile and kinesthetic sensory information. Contact image is constructed by guiding the sensory system through the environment using motor commands that depend on the objectives of the user. Given the large number of degrees of freedom, multiplicity of subsystems, spatial distribution of receptors and sensory-motor nature of haptic tasks, the human haptic capabilities and limitations that determine the characteristics of haptic devices are difficult to determine and characterize.

Haptic devices receive motor commands from the user and display the image of force distribution to the user. A haptic interface should provide a good match between the human haptic system and the device used for sensing and displaying haptic information. The primary input-output (measured and displayed) variables of the haptic interface are movement and force (or vice versa) with their spatial and temporal distributions. Haptic devices can therefore be treated as generators of mechanical impedance, which represents the relation between the force and movement (and their derivatives) in various positions and orientations. When displaying contact with a finite impedance, either force or movement represent excitation while the remaining quantity represents the response (if force is excitation then movement is response and vice versa), which depends on the implemented control algorithm. Consistency between the free movement of hands and touch is best achieved by taking into account the position and movement of hands as excitation and resultant vector of force and its distribution within the area of contact as response.

Since a human user senses and controls the position and force displayed by a haptic device, the performance specifications of the device directly depend on human capabilities. In many simple tasks that involve active touch, either tactile or kinesthetic information is of primary importance while the other is only complementary information. For example, when trying to determine the length of a rigid object by holding it between thumb and index finger, kinesthetic information is essential while tactile information is only supplementary. In this case, the crucial ability is sensing and controlling the position of the finger. On the other hand, perception of texture or slipperiness of the surface depends mainly on tactile information while kinesthetic information only supplements tactile perception. In this case, perceived information about temporal-spatial distribution of forces provides a basis for perceiving and inferring the conditions of contact and characteristics of the surface of the object. In more complex haptic tasks, however, both kinesthetic and tactile feedback are required for correct perception of the environment.

Due to hardware limitations, haptic interfaces provide stimuli that only approximate interaction with a real environment. However, this does not mean that an artificially synthesized haptic stimulus does not feel realistic. Consider the analogy with a visual experience of watching a movie. Although visual stimuli in the real world are continuous in time and space, visual displays project images with a frequency of only about 30 frames per second. Nevertheless, the sequence of images is perceived as a continuous scene since displays are able to exploit limitations of the human visual apparatus.

A similar reasoning also applies to haptic interfaces where implementation of appropriate situation-specific simplifications exploits limitations of the human haptic system. An understanding of human biomechanical, sensory-motor and cognitive capabilities is critical for proper design of device hardware and control algorithms for haptic interfaces.

Fig. 7.2 The term kinesthetics mainly refers to the perception of movement and position of limbs

7.1.1 Kinesthetic Perception

The term kinesthetics refers to the perception of movement and position of limbs and in a broader sense includes also perception of force. This perception originates primarily from mechanoreceptors in muscles, which provide the central nervous system with information about static muscle length, muscle contraction velocity and forces generated by muscles. Awareness of limb position in space, of limb movement and of mechanical properties (such as mass and stiffness) of objects with which the user interacts emerges from these signals. Sensory information about the change in limb position also originates from other senses, particularly from receptors in joints and skin. These senses are particularly important for kinesthetics of the arm. Receptors in the skin significantly contribute to the interpretation of the position and movement of the arm. The importance of cutaneous sensory information is not surprising considering the high density of mechanoreceptors in the skin and their specialization for tactile exploration. This feedback information is important for kinesthetics of the arm because of the complex anatomical layout of muscles that extend across a number of joints, which introduces uncertainty in the perception of position derived from receptors in muscles and tendons (Fig. 7.2).

Mechanoreceptors are primary and secondary receptors (also called Type Ia and Type II sensory fiber) located in muscle spindles. Muscle spindles are elongated structures 0.5–10 mm in length, consisting of muscle fiber bundles. Spindles are located parallel to the muscle fibers, which are generators of muscle force and are attached at both ends either to the muscle or tendon fibers [5]. A muscle spindle detects length and tension changes in muscle fibers. The main role of a muscle spindle is to respond to stretching of the muscle and to stimulate muscle contraction through a reflex arc to prevent further extension. Reflexes play an important role in the control of movement and balance. They allow automatic and rapid adaptation of muscles to changes in load and length.

Both primary and secondary spindle receptors respond to changes in muscle length. However, the primary receptors are much more sensitive to velocity and acceleration components of the movement and their response considerably increases with increased velocity of muscle stretching. The response of primary spindle receptors is nonlinear and their output signal depends on the length of the muscle, muscle contraction history, current velocity of muscle contraction and activity of the central nervous system, which modifies the sensitivity of muscle spindles. Secondary spindle receptors have a much less dynamic response and have a more constant output at a constant muscle length compared to the primary receptors. Higher dynamic sensitivity of primary spindle receptors indicates that these receptors mainly respond to the velocity and direction of muscle stretching or movement of a limb while the secondary spindle receptors measure static muscle length or position of the limb.

The second type of mechanoreceptors is a Golgi tendon organ. It measures 1 mm in length, has a diameter of 0.1 mm and is located at the attachment of a tendon to the bundle of muscle fibers. The receptor is therefore connected in series with the group of muscle fibers and primarily responds to the force generated by these fibers. When muscle is exposed to excessive load, the Golgi tendon organ becomes excited, which leads to the inhibition of motor neurons and finally to reduction of muscle tension. In this way, the Golgi tendon organ also serves as a safety mechanism that prevents damage to the muscles and tendons due to excessive loads.

Other mechanoreceptors found in joints are Ruffini endings, which are responsible for sensing angle and angular velocity of the joint movements, Pacinian corpuscles, which are responsible for estimation of the joint acceleration, and free nerve endings, which constitute the nociceptive system of the joint.

7.1.2 Tactile Perception

Although humans are presented with various sensations when touching objects, these sensations are a combination of only a few basic types of sensations, which can be represented with basic building blocks. Roughness, lateral skin stretch, relative tangential movement and vibrations are the basic building blocks of sensations when touching objects. Texture, shape, compliance and temperature are the basic object properties that are perceived by touch. Perception is based on mechanoreceptors in the skin. When designing a haptic device, human temporal and spatial sensory capabilities have to be considered (Fig. 7.3).

Four different types of sensory organs for sensing touch can be found in the skin. These are Meissner's corpuscles, Pacinian corpuscles, Merkel's discs and Ruffini corpuscles (Fig. 7.4). Figure 7.5 shows the rate of adaptation of these receptors to stimuli, the average size of the sensory area, spatial resolution, sensing frequency range and frequency of maximum sensitivity. Delays in the response of these receptors range from 50 to 500 ms.

Since the thresholds for different receptors overlap, the quality of sensing of touch is determined by a combination of responses of different receptors. Receptors complement each other, making it possible to achieve a wide sensing range for detecting vibrations with frequencies ranging from 0.4 to about 1,000 Hz [6, 7]. In general, the threshold for detecting tactile inputs decreases with increased duration

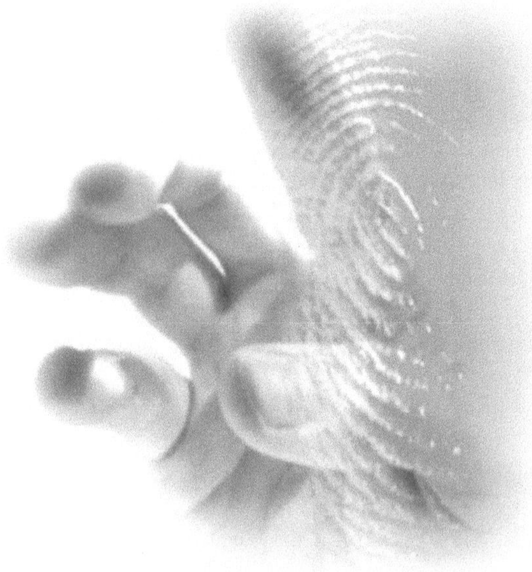

Fig. 7.3 The highest density of tactile receptors can be found in fingertips

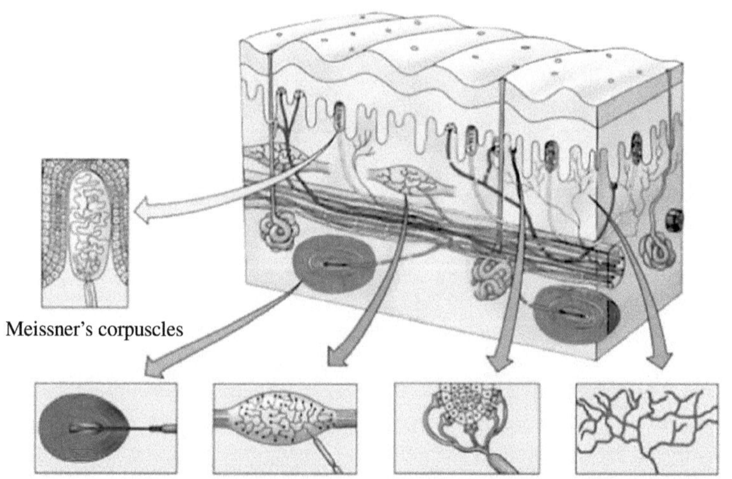

Fig. 7.4 Mechanoreceptors in the skin

of the stimuli. The spatial resolution at the fingertips is about 0.15 mm while the minimum distance between two points that can be perceived as separate points is approximately 1 mm. Humans can detect a 2-μm high needle on the smooth glass surface. Skin temperature affects the tactile perception.

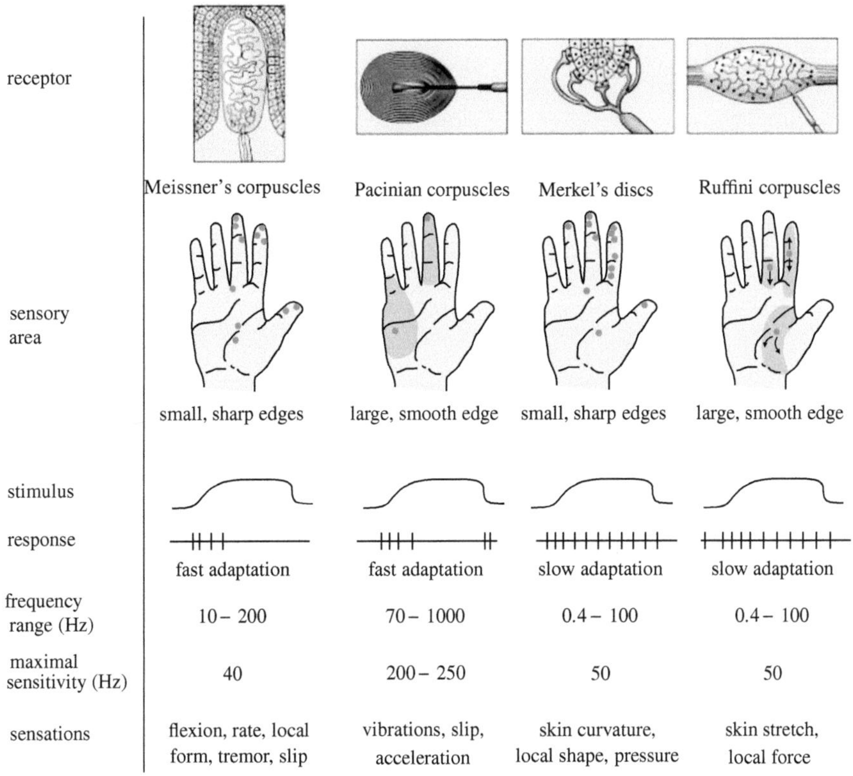

Fig. 7.5 Functional properties of skin mechanoreceptors

Properties of the human tactile perception provide important guidelines for planning and evaluation of tactile displays. The size of perception area, duration and frequency of the stimulus signal need to be considered.

7.1.3 Vestibular Perception

The vestibular system, which contributes to human balance and sense of spatial orientation, is the sensory system that provides the dominant input about the movement and equilibrioception. Together with the cochlea, a part of the auditory system, it constitutes the labyrinth of the inner ear (Fig. 7.6). As human movements consist of rotations and translations, the vestibular system comprises two components: the semicircular canal system, which indicates rotational movements; and the otoliths, which indicate linear acceleration. The vestibular system sends signals primarily to the neural structures that control eye movements and to the muscles that keep a body upright.

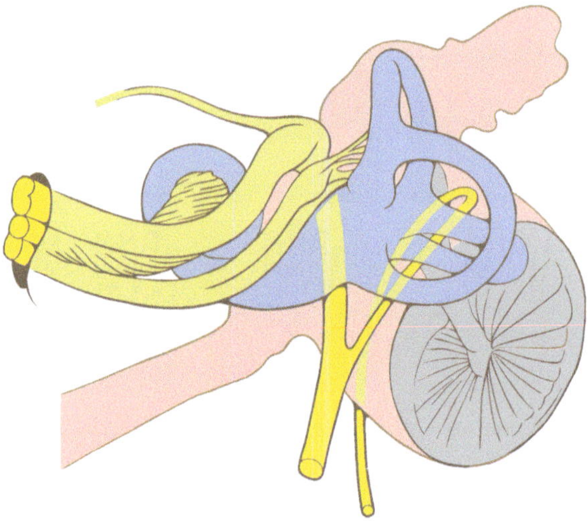

Fig. 7.6 Vestibular system, located in inner ear, contributes to human balance and sense of spatial orientation

7.1.4 Human Motor System

During haptic interactions, a user directly interacts with a haptic display through physical contact. As a consequences, this affects the stability of haptic interaction. It is therefore necessary to consider human motor properties to ensure stable haptic interaction.

The human arm is a complex biomechanical system whose properties cannot be uniquely described; it may behave as a system where position is controlled, or it may behave as a system where in a partly constrained movement the contact force is controlled.

A human arm can be modeled as a non-ideal source of force in interaction with a haptic interface. The term *non-ideal* in this case refers to the fact that the arm does not respond only to signals from the central nervous system, but also to the movements imposed by its interaction with the haptic interface. Relations are shown in Fig. 7.7.

Force $\mathbf{F}_h^\star$ is the component of the force resulting from muscle activity that is controlled by the central nervous system. If the arm does not move, the contact force $\mathbf{F}_h$ applied by the human arm on the haptic display equals $\mathbf{F}_h^\star$ (muscle force that initializes the movement). However, $\mathbf{F}_h$ is also a function of the imposed movement by the haptic display. If the arm moves (haptic display imposes movement), the force acting on the display differs from $\mathbf{F}_h^\star$. Conditions are presented in Fig. 7.7. Instantaneous force $\mathbf{F}_h$ is not only a function of the $\mathbf{F}_h^\star$ but also a function of movement velocity $\mathbf{v}_h$ of the contact point between the arm and the tip of the haptic interface. Considering the analogy between mechanical and electrical systems, force $\mathbf{F}_h$ can be written as

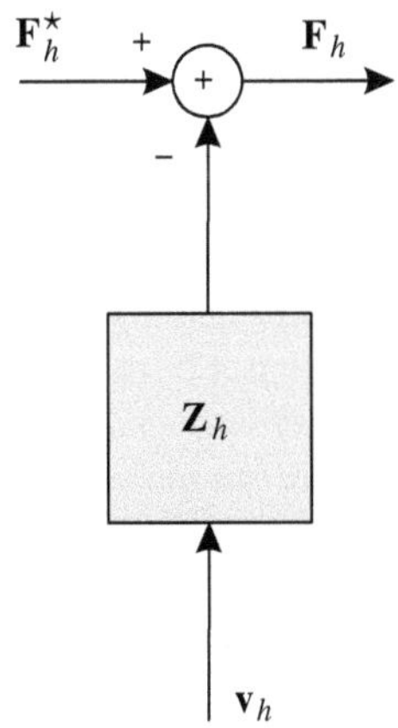

Fig. 7.7 Contact force $\mathbf{F}_h$ is a function of muscular force $\mathbf{F}_h^\star$ and arm impedance Z_h

$$\mathbf{F}_h = \mathbf{F}_h^\star - \mathbf{Z}_h \mathbf{v}_h, \tag{7.1}$$

where $\mathbf{Z}_h$ represents biomechanical impedance of the human arm and maps movement of the arm into force. $\mathbf{Z}_h$ is primarily determined by physical and neurological properties of the human arm and has an important role in the stability and performance of the haptic system.

7.2 Haptic Representation in Virtual Reality

Different characteristics of the real environment are perceived through the haptic sense. The objective of using haptic displays is to represent the virtual environment as realistically as possible. Abstract haptic representations are rarely used, except in interactions with scaled environments (e.g. nanomanipulation), for sensory substitution and for the purpose of avoiding dangerous situations. In interactions with scaled environments, the virtual reality application may use forces perceivable to humans, for example, to present events unfolding at the molecular level.

Information that can be displayed through haptic displays includes object features such as texture, temperature, shape, viscosity, friction, deformation, inertia and weight. Restrictions imposed by haptic displays usually prevent the use of combinations of different types of haptic displays.

In conjunction with visual and acoustic presentations, the haptic presentation is the one that the human cognitive system most relies on in the event of conflicting information.

Another important feature of haptic presentations is its local nature. Thus, it is necessary to haptically render only those objects that are in direct reach of the user. This applies only to haptic interactions since visual and auditory sensations can be perceived at a distance.

Haptic interaction is frequently used for exploration of objects in close proximity. Force displays are used in virtual reality for displaying object form, for pushing and deforming objects. Simulation of a virtual environment defines whether the applied force results in deflection or movement of objects. Haptic displays can be divided into three major groups.

Force displays are especially useful for interaction with a virtual environment, for the control or manipulation of objects and for precise operations.

Tactile displays are especially useful in cases where object details and surface texture are more important than overall form.

Passive haptic feedback can be based on the use of control props and platforms. These provide a passive form of haptic feedback.

7.2.1 Representational Models for Virtual Objects

Before dealing with methods for collision detection in virtual environments, we shall review basic concepts of geometric modeling of virtual objects, since the method for collision detection significantly depends on the object model [8–10]. Most methods for geometric modeling originate from computer graphics and were presented in Chap. 5.

Object models are often represented using the object's exterior surfaces—the problem of model representation is simplified to a mathematical model for describing the object's surface, which defines the outside boundaries of an object. These representations are often referred to as *representations with boundary surface*. Other representations are based on *constructive solid geometry*, where solid objects are used as basic blocks for modeling, or *volumetric representations*, which model objects with vector fields.

Haptic rendering is generally based on completely different requirements than computer graphics. The sampling frequency of a haptic system is significantly higher and haptic rendering is of a more local nature since we cannot physically interact with the entire virtual environment at once. Haptic rendering thus constructs a specific set of techniques making use of representational models developed primarily for computer graphics.

The following section provides an overview of some modeling techniques for virtual objects with an emphasis on attributes specific for haptic collision detection.

Two early approaches for representation of virtual objects were based on a force vector field method and an intermediate plane method. The vector field corresponds to the desired reaction forces. The interior of an object is divided into areas whose main characteristic is the common direction of force vectors, whereas the force vector length is proportional to the distance from the surface (Fig. 7.8).

An intermediate plane [11], on the other hand, simplifies representation of objects modeled with boundary surfaces. The intermediate plane represents an approximation of the underlying object geometry with a simple planar surface. The plane parameters are refreshed as the virtual tool moves across the virtual object. However, the refresh rate of the intermediate plane can be lower than the frequency of the haptic system.

Other representational models originate from the field of computer graphics. Two frequently used representations are implicit and parametric surfaces.

An implicit surface is defined by the implicit function. It is defined by mapping the three-dimensional space to the space of real numbers $f : \Re^3 \mapsto \Re$ and an implicit surface is defined with points, where $f(x, y, z) = 0$. Such a function uniquely defines what is inside $f(x, y, z) < 0$ and what is outside $f(x, y, z) > 0$ of the model. Implicit surfaces are consequently generically closed surfaces.

A parametric surface is defined by mapping from a subset of the plane into a three-dimensional space $f : \Re^2 \mapsto \Re^3$. Contrary to implicit surfaces, parametric surfaces are not generically closed surfaces. They thus do not present the entire object model, but only a part of the object boundary surface. Implicit and parametric surfaces were presented in more details in Sect. 5.2.2.

However, a representational method that is most often used in computer graphics is based on polygonal models. Representations using polygons are simple, polygons are versatile and appropriate for fast geometric computations. Polygon models enable presentation of objects with boundary surfaces. An example of a polygonal model is shown in Fig. 7.9, where the most simple polygons—triangles—are used. Each object surface is represented with a triangle that is defined with three points (for example, $tr_1 = \triangle P_0 P_1 P_2$).

Haptic rendering based on polygonal models may cause force discontinuities at the edges of individual polygons, when the normal of the force vector moves from the current to the next polygon. The human sensing system is accurate enough to perceive such discontinuities, so they must be compensated for. A method for removing discontinuities is referred to as *force shading* and is based on interpolation of normal vectors between two adjacent polygons.

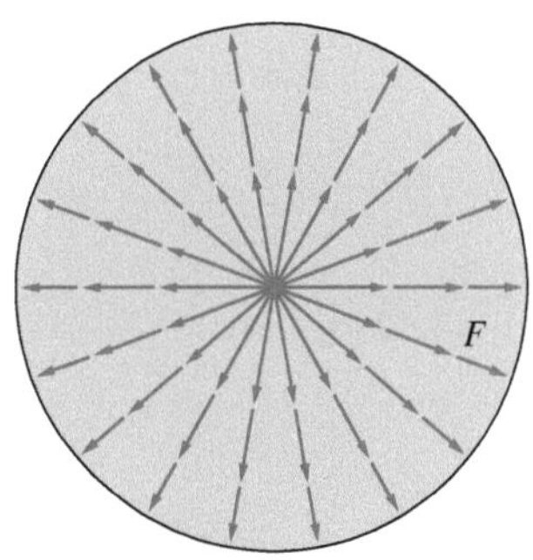

Fig. 7.8 Spherical object modeling using a force vector field method

7.3 Collision Detection

The algorithm for haptic interaction with a virtual environment consists of a sequence of two tasks. When the user operates a virtual tool attached to a haptic interface, the new tool pose is computed and possible collisions with objects in a virtual environment are determined. In case of contact, reaction forces are computed based on the environment model and force feedback is provided to the user via the haptic display. Collision detection guarantees that objects do not float into each other. A special case of contact is the grasping of virtual objects that allows object manipulation (Fig. 7.10). If grasping is not adequately modeled, it might happen that the virtual hand may pass through the virtual object and the reaction forces that the user perceives are not consistent with the visual information.

If virtual objects fly through each other, this creates a confusing visual effect Penetration of one object into another thus needs to be prevented. When two objects try to penetrate each other, we are dealing with collision.

Collision detection is an important step toward physical modeling of a virtual environment. It includes automatic detection of interactions between objects and computation of contact coordinates. At the moment of collision, the simulation generates a response to the contact. If the user is coupled to one of the virtual objects (for example, via a virtual hand), the response to the collision results in forces, vibrations or other haptic quantities being transmitted to a user via a haptic interface.

7.3.1 Collision Detection and Object Deformation

Haptic rendering represents computation of forces required for generation of impression of a contact with the virtual object. These forces typically depend on the penetration depth of the virtual tool into the virtual object and on the direction of the tool acting on the object. Due to the complexity of computation of reaction forces in the

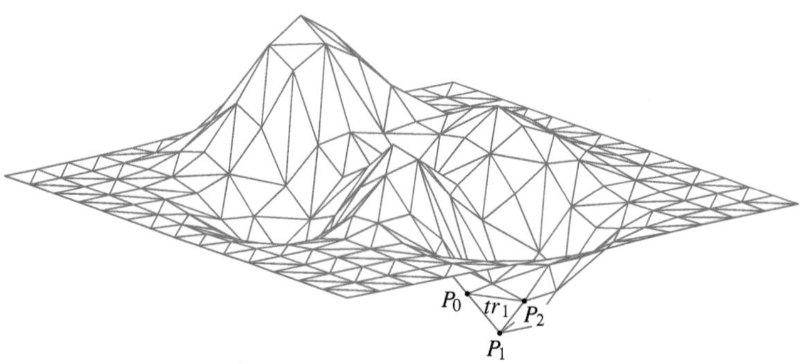

Fig. 7.9 Object modeling using triangles

case of complex environments, a virtual tool is often simplified to a point or a set of points representing the tool's endpoint. The computation of penetration depth thus simplifies to finding the point on the object that is the closest to the tool endpoint. As the tool moves, the closest point on the object also changes and must be refreshed with the sampling frequency of the haptic system.

We will analyze a simple single-point haptic interaction of a user with a virtual environment. In this case, the pose of the haptic interface end-effector is measured. This is referred to as a *haptic interaction point* **HIP**. It is then necessary to determine if the point lies within the boundaries of the virtual object. If this is the case, then the penetration depth is computed as the difference between **HIP** and the corresponding point on the object's surface (*surface contact point* **SCP**). Finally, the resulting reaction force is estimated based on the physical model of the virtual object [12]. Figure 7.11 illustrates relations during a haptic contact in a virtual environment. Points $\mathbf{P}_1$ and $\mathbf{P}_2$ define a vector $\overrightarrow{\mathbf{P}_1\mathbf{P}_2}$, that determines the intersecting ray with a virtual object; in other words, a ray that penetrates the object's surface. Point $\mathbf{P}_1$ is usually defined by the pose of the tool at the moment just before touching the object while point $\mathbf{P}_2$ is defined by the pose of the tool after touching the object in the next computational step. Point **HIP** equals point $\mathbf{P}_2$.

A more detailed analysis of methods for collision detection between objects in a virtual environment was presented in Chap. 3.

Fig. 7.10 Grasping of virtual objects

7.4 Haptic Rendering in Virtual Reality

Computer haptics is a research area dealing with techniques and processes related to generation and rendering of contact properties in a virtual environment and displaying of this information to a human user via a haptic interface. Computer haptics deals with models and properties of virtual objects as well as algorithms for displaying haptic feedback in real time.

Haptic cue rendering often represents the most challenging problem in a virtual reality system. The reason is primarily the direct physical interaction and, therefore, a bidirectional communication between the user and the virtual environment through a haptic display. The haptic interface is a device that enables man-machine interaction. It simultaneously generates and perceives mechanical stimuli.

Haptic rendering allows the user to perceive the mechanical impedance, shape, texture and temperature of objects. When pressing on an object, the object deforms due to its final stiffness or moves if it is not grounded. The haptic rendering method must take into account the fact that humans simultaneously perceive tactile as well as kinesthetic cues. Due to the complexity of displaying tactile and kinesthetic cues, virtual reality systems are usually limited to only one type of cue. Haptic rendering can thus be divided into rendering through the skin (temperature and texture) and rendering through muscles, tendons and joints (position, velocity, acceleration, force and impedance).

Stimuli that mainly trigger skin receptors (e.g. temperature, pressure, electrical stimuli and surface texture) are displayed through tactile displays. Kinesthetic information that enables the user to investigate object properties such as shape, impedance (stiffness, damping, inertia), weight and mobility, is usually displayed through robot-based haptic displays. Haptic rendering can produce different kinds of stimuli, ranging from heat to vibrations, movement and force. Each of these stimuli must be rendered in a specific way and displayed through a specific display.

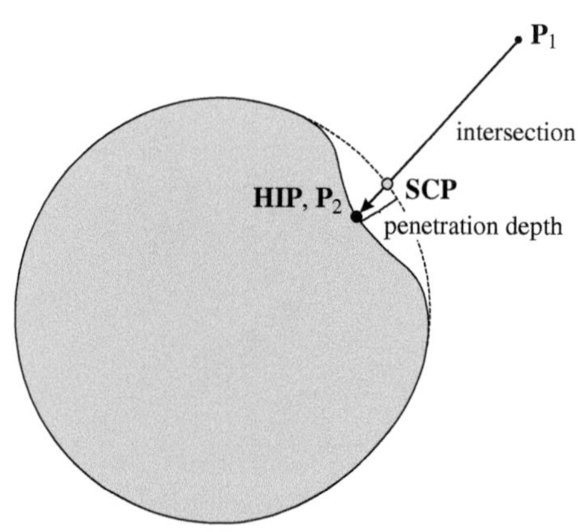

Fig. 7.11 Conditions during a penetration into the virtual object

Haptic rendering with low sampling frequency or high latency may influence the perception of a virtual environment and may cause instability of the haptic display. This is completely different from visual rendering, where slow processing causes the user to perceive visual information not as a continuous stream but as a discrete sequence of images. However, each image is still a faithful representation of a virtual environment at a given time. For example, visual representation of a brick would still display a brick while the haptic system would render it as a mass of clay due to the low stiffness of the virtual object, which is a result of the low sampling frequency.

7.4.1 Rendering of Temperature, Texture and Force

Visual rendering results in a visual image that is transmitted to the user from the display via electromagnetic radiation. Haptic rendering, on the other hand, enables implementation of different types of stimuli, from vibration to movement and force. Each stimulus is rendered in a specific manner and presented through a specific display.

Temperature rendering is based on heat transfer between the display and the skin. The tactile display creates a sense of object temperature.

Texture rendering provides tactile information and can be achieved, for example, using a field of needles that simulates the surface texture of an object. Needles are active and adapt according to the current texture of the object being explored by the user.

Kinesthetic rendering allows display of kinesthetic information and is usually based on the use of robots. By moving the robot end-effector, the user is able to haptically explore his surroundings and perceive the position of an object. The object is determined by the inability to penetrate the space occupied by that particular entity. The greater the stiffness of the virtual object, the stiffer the robot manipulator becomes while in contact with a virtual object. Kinesthetic rendering thus enables perception of the object's mechanical impedance.

Haptic rendering of a complex scene is much more challenging compared to visual rendering of the same scene. Therefore, haptic rendering is often limited to simple virtual environments. The complexity arises from the need for a high sampling frequency in order to provide consistent feeling of rendered objects. If the sampling frequency is low, the time required for the system to respond and produce an adequate stiffness (for example, during penetration into a virtual object) becomes noticeable. Stiff objects consequently feel compliant.

The complexity of realistic haptic rendering depends on the type of simulated physical contact implemented in the virtual reality. If only the shape of an object is being displayed, then *touching* the virtual environment with a pencil-style probe is sufficient. Substantially more information needs to be transmitted to the user if it is necessary to grasp the object and raise it to feel its weight, elasticity and texture. The form of the user contact with the virtual object thus needs to be taken into account for haptic rendering (for example, contact can occur at a single point, the object can be grasped with the entire hand or with a pinch grip between two fingers).

Single-point contact is the most common method of interaction with virtual objects. The force display provides stimuli to a fingertip or a probe that the user holds with his fingers. The probe is usually attached as a tool at the tip of the haptic interface. In the case of the single point contact, rendering is usually limited to the contact forces only and not contact torques.

Two-point contact (*pinch grip*) enables display of contact torques through the force display. With a combination of two displays with three degrees of freedom it is, in addition to contact forces, possible to simulate torques around the center point on the line, which connects the points of touch.

Multipoint contact allows object manipulation with six degrees of freedom. The user is able to modify both the position and the orientation of the manipulated object. To ensure adequate haptic information, it is necessary to use a device that covers the entire hand (a haptic glove).

As with visual and acoustic rendering, the amount of details or information that can be displayed with haptic rendering is limited. The entire environment usually needs to be displayed in a haptic form. However, due to the complexity of haptic rendering algorithms and the specificity of haptic sensing, which is local in nature, haptic interactions are often limited to contact between the probe and a small number of nearby objects. Due to the large amount of information necessary for proper representation of object surfaces and dynamic properties of the environment, haptic rendering requires a more detailed model of a virtual environment (object dimensions, shape and mechanical impedance, texture, temperature) than is required for visual or acoustic rendering. Additionally, haptic rendering is computationally more demanding than visual rendering since it requires accurate computation of contacts between objects or contacts between objects and tools or avatars. These contacts form the basis for determining reaction forces.

Haptic interfaces provide haptic feedback about the computer-generated or remote environment to a user who interacts with this environment. Since these interfaces do not have their own intelligence, they only allow presentation of computer-generated quantities. For this purpose it is necessary to understand physical models of a virtual environment that enable generation of time-dependent variables (forces, accelerations, vibrations, temperature, …) required for control of the haptic interface.

7.4.2 Rendering of Simple Contact

The task of *haptic rendering* is to enable the user to touch, sense and manipulate virtual objects in a simulated environment via a haptic interface [13, 14]. The basic idea of haptic rendering can be explained using Fig. 7.12, where a frictionless sphere is positioned in the origin of a virtual environment. Now assume that the user interacts with a sphere in a single point, which is defined by the haptic interface end-effector position (**HIP**) [15]. In the real world this would be analogous to touching a sphere with a tip of a thin stick. When moving through a free space, the haptic interface behaves *passively* and does not apply any force onto the user until the occurrence of contact with a sphere. Since the sphere has finite stiffness, the **HIP** penetrates into the sphere at the point of contact. When contact with the sphere is detected, the corresponding reaction force is computed and transmitted via a haptic interface to the user. The haptic interface becomes *active* and generates a reaction force that prevents further penetration into the object. The magnitude of the reaction force can be computed based on a simple assumption that the force is proportional to the penetration depth. With the assumption of a frictionless sphere, the reaction force direction is determined as a vector normal to the sphere surface at the point of contact.

7.4.3 Stiffness and Compliance Models

In general, two models of haptic interaction with the environment can be distinguished. The first model is called a *compliance model* while the second is called a *stiffness model*. The two terms refer to a simple elastic model $F = Kx$ of a wall with stiffness K, penetration depth x and reaction force F. The two concepts for modeling haptic interaction with the environment are shown in Figs. 7.13 and 7.14.

- In the case of the stiffness model in Fig. 7.13, the haptic interface measures displacement x and the simulation returns the corresponding force F as a $F = Kx$. Haptic interfaces that are excellent force sources are suitable for implementation of such a model.

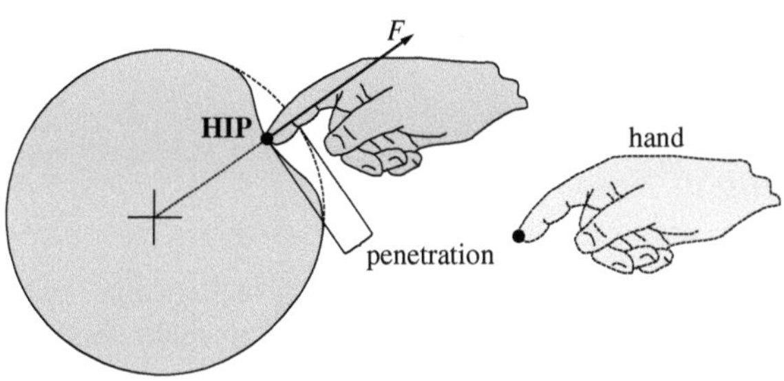

Fig. 7.12 Haptic rendering of a sphere in a virtual environment

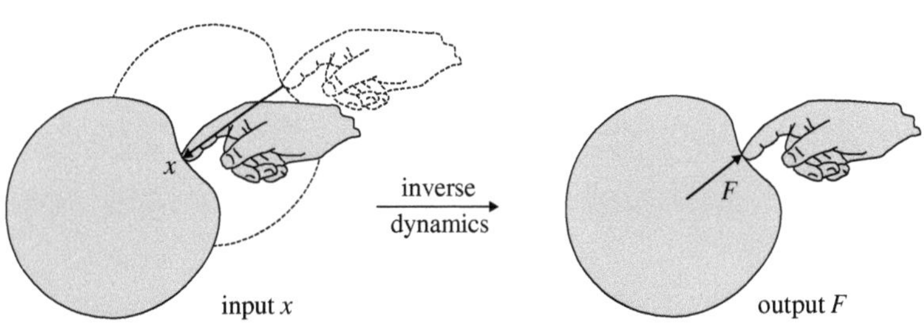

Fig. 7.13 Stiffness (impedance) model of haptic interaction

- In the case of the compliance model in Fig. 7.14, the haptic interface measures the force F between the user and the haptic display and the simulation returns the displacement x as a result of relation $x = K^{-1}F = CF$, where compliance C is defined as the inverse value of stiffness K. Stiff haptic interfaces, such as industrial manipulators equipped with a force and torque sensor, are suitable for implementation of such a model.

In the case of more complex models, where viscous and inertial components are present in addition to compliance, the terms stiffness model and compliance model are substituted with the terms *impedance model* (an equivalent of the stiffness model) and *admittance model* (an equivalent of the compliance model). For the purpose of generality we will use the terms impedance and admittance in the following sections. When an object is displaced due to contact, object dynamics need to be considered to determine the relation between force and displacement. An inverse dynamic model is required for computing the impedance and a forward dynamic model is required for computing the admittance causality structure.

Most approaches to haptic rendering are based on the assumption of interactions with stiff grounded objects, where stiffness characteristics dominate over other dynamic properties. In the case of objects in a virtual or remote environment that are displaced as a result of haptic interactions, the object deformation is masked with the object displacement. In such cases it is reasonable to attribute the entire haptic interface displacement to the object movement without considering the object deformation. Namely, we can assume that the majority of real objects do not deform considerably under the contact forces. However, such an assumption is not valid when the impedance, which is the result of object displacement, is approximately equal to the impedance properties for object deformation.

7.4.4 Modeling of Free Space

A system for haptic rendering must be capable of realistically rendering motion in a free space (without interaction forces). In order to make this possible, a haptic

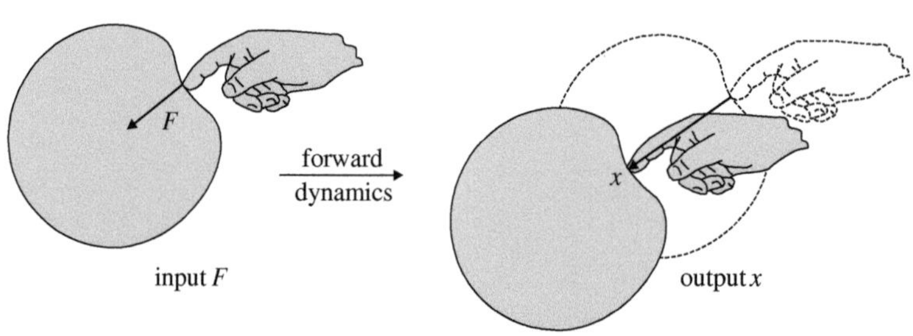

Fig. 7.14 Compliance (admittance) model of haptic interaction

interface must possess characteristics that enable it to be moved without significant physical effort and with the smallest possible disturbance resulting from friction in the mechanism, intrinsic inertia and vibrations. This can be achieved passively with a mechanism design based on low intrinsic impedance (small inertia, damping and elasticity), or actively, with the control system compensating the device's intrinsic impedance.

7.4.5 Modeling of Object Stiffness

The maximal object stiffness that can be realistically rendered and displayed represents a critical characteristic of a haptic display and control algorithm. Haptic interface bandwidth, stiffness, resolution and the capability of generating forces are quantities that decisively affect accuracy of displaying transient responses during contacts with virtual objects. In addition, the contact force modeling approach affects contact intensity and system stability.

At the moment of collision with the surface of a virtual object, an instantaneous change of force occurs with an impulse large enough to withdraw the momentum from the user's hand or a tool. This requires high-frequency bandwidth that enables fast and stable change of force. On the other hand, continuous contact with the object surface requires large haptic interface end-effector forces without saturation of device actuators and unstable behavior. Accurate transient responses and stable steady-state behavior enable realistic impression of object stiffness.

The most common way of modeling stiff and grounded surface is based on a model consisting of a parallel connection of a spring with stiffness K and a damper with viscosity B, as shown in Fig. 7.15. Given $x = 0$ for an undeformed object and $x < 0$ inside the object boundaries, the modeled contact force on the virtual hand as shown in Fig. 7.15 equals

$$F = \begin{cases} -(Kx + B\dot{x}) & \text{for } x < 0 \wedge \dot{x} < 0 \\ -Kx & \text{for } x < 0 \wedge \dot{x} \geq 0 \\ 0 & \text{for } x \geq 0. \end{cases} \tag{7.2}$$

In this case, the viscous damping behaves as a directed damper that is active during the penetration into the object and passive during the withdrawal from the object. This enables a stable and damped contact with the object as well as a realistic contact rendering. Contact relations are shown in Fig. 7.16 for a one degree of freedom system. From relations shown in Fig. 7.16 it is apparent that at the instance of the contact, there is a step change in the force signal due to the contribution of viscous damping, since the approach velocity differs from zero. During the penetration into the object the influence of the viscous part is being reduced due to the decreasing movement velocity. At the same time the contribution of the elastic element increases due to the increased depth of penetration into the object. At the instance of the largest object deformation, the penetration velocity equals zero and the reaction force is only the result of the elastic element. Since the damper operates in a single direction (only

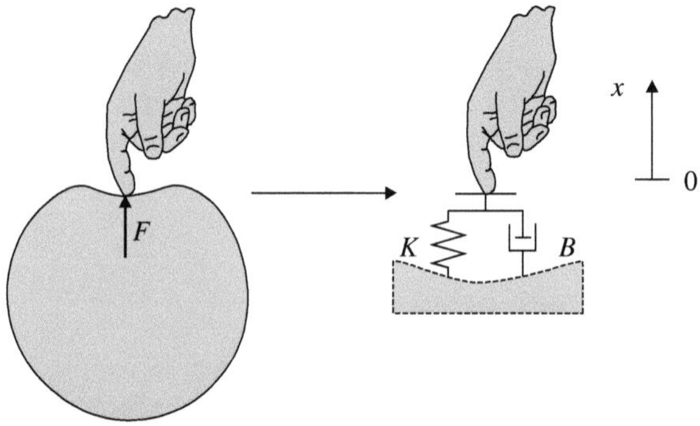

Fig. 7.15 Spring-directed damper model of a contact with a grounded object

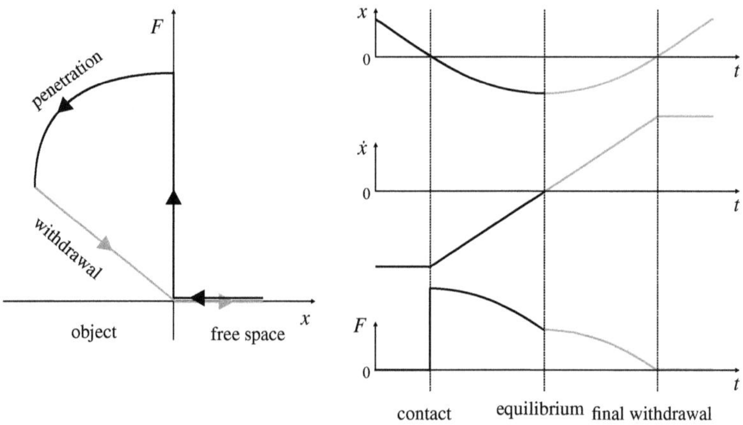

Fig. 7.16 Relations during a contact simulation with a spring-directed damper model

active during penetration and inactive during withdrawal), this results in a linearly decreasing reaction force as a function of displacement x. The reaction force reaches zero at the boundary of the undeformed object, resulting in a smooth transition between the object and free space. Such a modeling approach guarantees pronounced initial contact, rigidity of a stiff surface and a smooth withdrawal from the object surface.

7.5 Control of Haptic Interfaces

Forces and movements that are computed based on dynamic model of the virtual environment can be used as an input signal to the controller of the haptic interface. Selection of control strategy (impedance or admittance control) depends on the available hardware and software architectures as well as on the planned use of the haptic interface.

Interaction between the user and the environment presents a bilateral transfer of energy, as the product of force and displacement defines the mechanical work. The rate of change of energy (mechanical power) is defined by the instantaneous product of the interaction force and the movement velocity. The exchange of mechanical energy between the user and the haptic interface is the main difference compared to other display modalities (visual, acoustic) that are based on one-way flow of information with negligible energy levels.

If energy flow is not properly controlled, the effect of haptic feedback can be degraded due to unstable behavior of the haptic device. Important issues related to control of haptic interaction include its quality and especially stability of haptic interaction while taking into account properties of the human operator, who is inserted into the control loop [16, 17].

In the chapter related to modeling of collision detection we introduced concepts of impedance and admittance models of a virtual environments. Similarly, two classes of control schemes for control of haptic interfaces can be defined: (1) impedance control, which provides force feedback and (2) admittance control, which provides displacement feedback.

The impedance approach to displaying kinesthetic information is based on measuring the user's motion velocity or limb position and implementation of a force vector at the point of measurement of position or velocity. We will assume that the point of interaction is the user's arm. Even though it is also possible to construct kinesthetic displays for other parts of the body, the arm is the primary human mechanism for precise manipulation tasks. The magnitude of the displayed force is determined as a response of a simulated object to displacement measured on the user's side of the haptic interface.

Figure 7.17 shows a block scheme of impedance-controlled haptic interface. Joint position encoders measure angular displacements $\mathbf{q}^*$. These are then used in the forward kinematic model to compute the pose of the haptic interface end-effector $\mathbf{x}^*$. Desired reaction forces $\mathbf{F}_e$ are computed based on the physical model of the environment and the haptic interface end-effector pose (interaction force between the user and the interface can be used as an additional input). The desired force is then transformed into the desired joint torques through the manipulator-transposed Jacobian matrix, and haptic display actuators are used to generate the desired torques. Actuator torques result in a haptic display end-effector force that is perceived by the user. Thus, the haptic interface generates forces resulting from interactions in the virtual environment.

The main characteristics of an impedance display can be summarized as [19]:

- it has to enable unobstructed movement of the user arm when there is no contact with the environment,
- it has to exactly reproduce forces that need to be applied on the user,
- it has to generate large forces in order to simulate or reproduce contacts with stiff objects and

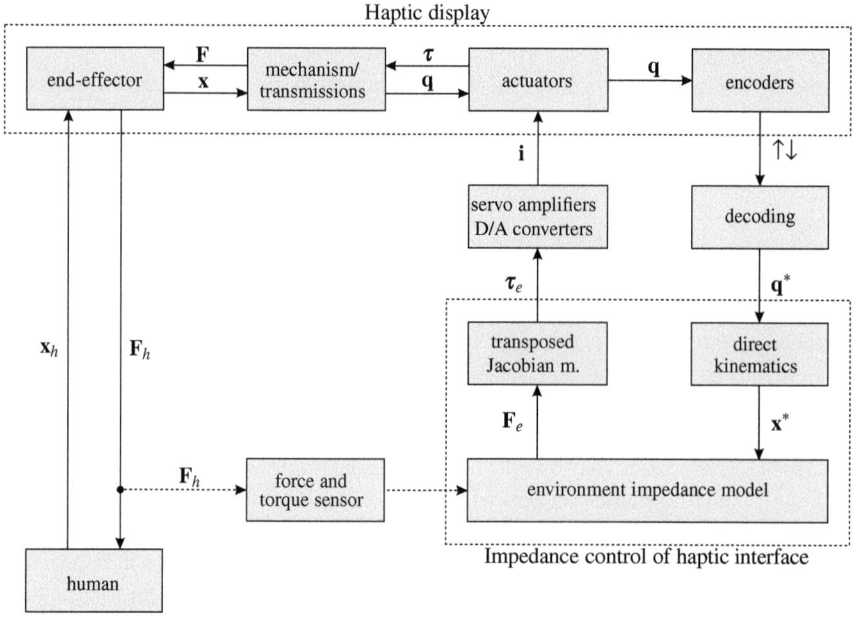

Fig. 7.17 Block scheme of an impedance controlled haptic interface. *Arrows* indicate dominant direction of flow of information. The *hatched line* indicates supplementary information. Adapted from [18]

- it has to have a wide bandwidth in order to reproduce transient responses during collisions with objects with sufficient fidelity and accuracy.

The above requirements are similar to the requirements for robots that are in contact with the environment and where interaction forces between the robot and the environment need to be precisely controlled.

An opposite approach to controlling haptic interfaces is based on measurement of forces, which the user applies on the haptic interface, and generation of displacements through the programmed motion of the haptic interface. The applied displacement is computed based on a dynamic model of the environment as a response of the environment to the measured force applied by the user.

Figure 7.18 shows the block scheme of the admittance-controlled haptic interface. Joint encoders measure displacements $\mathbf{q}^*$. These are then used in the forward kinematic model of the haptic display for computation of end-effector pose $\mathbf{x}^*$. Based on the physical model of the environment and the force being applied by the user on the robot end-effector (information about the haptic interface pose can be used as supplementary information), the desired haptic interface end-effector displacements $\mathbf{x}_e$ can be determined. These are then used as a reference input for the device position controller, whose output are the desired joint torques that must be generated by the haptic interface actuators. These generate movement of the haptic display

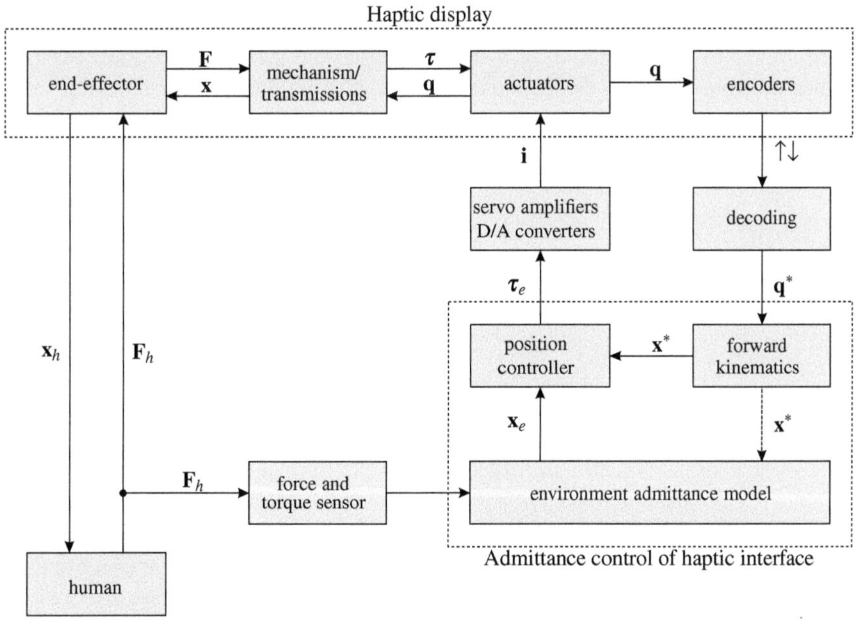

Fig. 7.18 Block scheme of an admittance controlled haptic interface. *Arrows* indicate dominant direction of flow of information. The *hatched line* indicates supplementary information

that is finally perceived by the human operator. Thus, a haptic interface displays displacements resulting from interactions with the virtual environment.

The main characteristics of an admittance display can be summarized as [19]:

- the mechanism needs to be stiff enough to completely prevent movement of the user's arm when in contact with a stiff object,
- it has to exactly reproduce desired displacement,
- it has to be backdrivable to allow reproduction of free movement and
- bandwidth of the system needs to be large enough to allow reproduction of transient responses with sufficient fidelity and accuracy.

The above characteristics are similar to the characteristics of position-controlled robot manipulators, where high accuracy of positional tracking is required.

In some cases the interaction force can be used as an additional input to the impedance controller. The displacement can be used as a supplementary input for the admittance controller. The class of the control scheme is therefore usually defined based on the output of the haptic interface (force, displacement). Impedance control is usually implemented in systems where the simulated environment is highly compliant while the admittance control approach is usually used in scenarios where the environment is very stiff. Selection of the type of controller does not depend only on the type of environment, but also on the dynamic properties of a haptic display. In the case of a haptic display with low impedance, where a force sensor is rarely mounted on the device end-effector, use of an impedance-type controller is more appropriate. On the other hand, an admittance controller is more appropriate for haptic displays that are based on industrial manipulators with high impedances, a force sensor attached to the end-effector and the capability of producing larger forces and torques.

7.6 Haptic Displays

A haptic display is a mechanical device designed for transfer of kinesthetic or tactile stimuli to the user. Haptic displays differ in their kinematic structure, workspace and output force (Fig. 7.19). In general, they can be divided into devices that *measure movement and display force* and devices that *measure force and display movement*. The former are called *impedance displays* while the latter are called *admittance displays*. Impedance displays typically have small inertia and are backdrivable. Admittance displays typically have much higher inertia, are not backdrivable and are equipped with a force and torque sensor. *A haptic interface* comprises everything between the human and the virtual environment. A haptic interface always includes a haptic display, control software and power electronics. It may also include a virtual coupling that connects the haptic display to the virtual environment. The haptic interface enables exchange of energy between the user and the virtual environment and is therefore important in the analysis of stability as well as efficiency.

The sense of touch is often used to verify object existence and mechanical properties. Since the haptic sense is difficult to deceive, implementation of a haptic display is very challenging. In general, the use of haptic displays in virtual reality applications is less frequent than the use of visual and sound displays. Haptic displays are more difficult to implement than visual or audio displays due to the bidirectional nature of the haptic system. Haptic displays do not only enable perception of the environment, but also manipulation of objects in the environment. Thus, the display requires direct contact with the user.

7.6.1 Kinesthetic Haptic Displays

Haptic displays are devices composed of mechanical parts, working in physical contact with a human body for the purpose of exchanging information. When executing tasks with a haptic interface, the user transmits motor commands by physically manipulating the haptic display, which displays a haptic sensory image to the user in the opposite direction via correct stimulation of tactile and kinesthetic sensory systems. This means that haptic displays have two basic functions: (1) to measure positions and interaction forces (and their time derivatives) of the user's limb (and/or other parts of the human body) and (2) display interaction forces and positions (and their spatial and temporal distributions) to the user. The choice of the quantity (position or force) that defines motor activity (excitation) and haptic feedback (response)

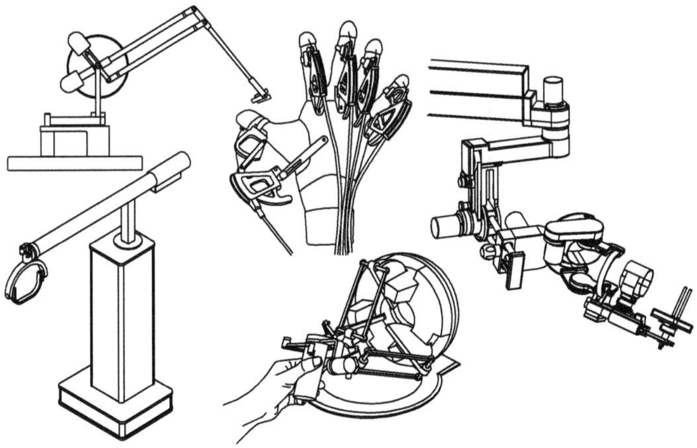

Fig. 7.19 A collage of different haptic robots for the upper extremities: Phantom (Sensable), Omega (Force dimension), HapticMaster (Moog FCS), ARMin (ETH Zurich) and CyberGrasp (CyberGlove systems)

depends on the hardware and software implementation of the haptic interface as well as on the task for which the haptic interface is used [20–22].

A haptic display must satisfy at least a minimal set of kinematic, dynamic and ergonomic requirements in order to guarantee adequate physical efficiency and performance with respect to the interaction with a human operator.

7.6.1.1 Kinematic Properties of Haptic Displays

A haptic display must be capable of exchanging energy with the user across mechanical quantities, such as *force* and *velocity*. The fact that both quantities exist simultaneously on the user side as well as on the haptic display side means that the haptic display mechanism must enable a continuous contact with the user for the whole time, when the contact point between the user and the device moves in a three-dimensional space.

The most important kinematic parameter of a haptic display is the number of degrees of freedom. In general, the higher the number of degrees of freedom, the greater the number of directions in which it is possible to simultaneously apply or measure forces and velocities. The number of degrees of freedom, the type of degrees of freedom (rotational or translational joints) and the length of the segments determine the workspace of the haptic display. In principle, this should include at least a subset of the workspace of human limbs, but its size primarily depends on the tasks for which it is designed.

An important aspect of kinematics of a haptic display presents the analysis of singularities [19]. The mechanism of the display becomes singular when one or more joints are located at the limits of their range of motion or when two or more joint axes become collinear. In a singular pose, the mechanism loses one or more of its degrees of freedom.

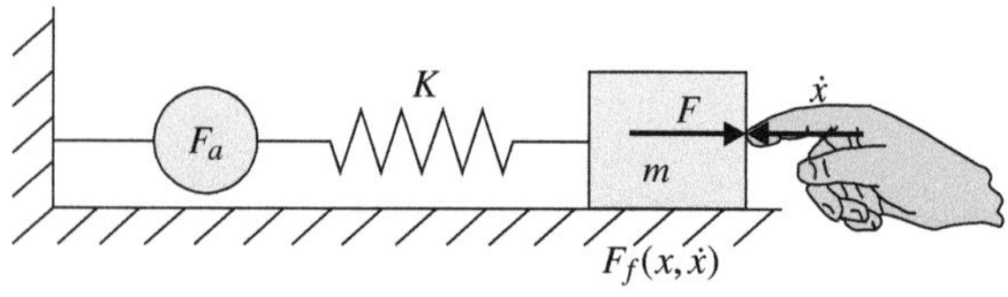

Fig. 7.20 Dynamic model of a haptic display with a single degree of freedom (adapted from [19])

7.6.1.2 Dynamic Properties of Haptic Displays

The intrinsic haptic display dynamics distorts forces and velocities that should be displayed to the user. A convincing presentation of contact with a stiff object, for example, requires high frequency response bandwidth of a haptic system. Thus, persuasiveness of force and velocity rendering is limited by the intrinsic dynamics of the haptic display. The effect of the intrinsic dynamics can be analyzed in a case study with a simplified haptic device consisting of a single degree of freedom as shown in Fig. 7.20 [19]. A haptic display applies force on the user while the user determines the movement velocity. An ideal display would allow undistorted transfer of a desired force ($F = F_a$; F_a is the actuator force and F is the force applied on the user) and precise velocity measurement ($\dot{x}_m = \dot{x}$; $\dot{x}$ is the actual velocity of the system endpoint and $\dot{x}_m$ is the measured velocity of the system endpoint). However, by taking into account the haptic display dynamics, the actual force applied on the user equals

$$F = F_a - F_f(x, \dot{x}) - m\ddot{x}\,. \tag{7.3}$$

Thus, the force perceived by the user is reduced by the effect of friction $F_f(x, \dot{x})$ and inertia m of the haptic display. In this simplified example the stiffness K does not affect the transfer of forces.

Equation (7.3) indicates that the mass of the haptic display affects the transmission of force to the user by resisting the change of velocity. This opposing force is proportional to the acceleration of the display. Minimization of the haptic display mass is necessary, since during collisions with virtual objects large accelerations (decelerations) can be expected. In case of multidimensional displays the dynamics becomes more complex. Except in specific cases where the dynamics of the mechanism is uncoupled (Cartesian mechanism), in addition to inertia, also Coriolis and centripetal effects cause absorption of actuation forces at velocities different from zero. A haptic display must be able to support its own weight in the gravitational field, as otherwise the gravitational force that is not associated with the task is transferred to the user. Gravity compensation can be achieved either actively through the display's actuators or passively with counterweights, which further increase inertia of the display.

Equation (7.3) indicates that part of the forces being generated by the actuators are absorbed due to friction. Friction occurs when two surfaces that are in physical contact move against each other. In general, friction can be separated into three components: static friction (a force that is required to initiate the motion between two surfaces, one against the other), Coulomb friction, which is velocity independent, and viscous damping, which is proportional.

7.6.2 Tactile Haptic Displays

Kinesthetic haptic displays are suitable for relatively coarse interactions with virtual objects, but tactile displays must be used for precise rendering. Tactile sensing plays an important role during manipulation and discrimination of objects, where force sensing is not efficient enough. Sensations are important for assessment of local shape, texture and temperature of objects as well as for detecting slippage. Tactile senses also provide information about compliance, elasticity and viscosity of objects. Vibrations sensing is important for detection of the objects' textures as well as for measuring vibrations. At the same time it also shortens reaction times and minimizes contact forces. Since reaction force is not generated prior to object deformation, tactile information also becomes relevant for the initial contact detection. This significantly increases abilities for detecting contacts, measuring contact forces and tracking a constant contact force. Finally, tactile information is also necessary for minimizing interaction forces in tasks that require precise manipulations.

In certain circumstances a tactile display of one type can be replaced with a display of another type. A temperature display can, for example, be used for simulating object material properties.

Tactile stimulation can be achieved using different approaches. Systems that are most often used in virtual environments include mechanical needles actuated using electromagnets, piezoelectric crystals or shape-memory alloy materials, vibrators that are based on sound coils, pressure from pneumatic systems or heat pumps.

7.6.3 Vestibular Displays

The vestibular sense enables control of balance. The vestibular receptor is located in the inner ear. It senses acceleration and orientation of the head in relation to the gravity vector. The relation between vestibular sense and vision is very strong and the discrepancy between the two inputs can lead to nausea.

The vestibular display is based on the user's physical movement. A movement platform can move the ground or the seat of the user. Such platforms are typical in flight simulators. A vestibular display alone cannot generate a convincing experience, but can be very effective in combination with visual and audio displays.

7.6.4 Characteristics of Haptic Displays

Haptic interactions that affect the design of haptic displays can be divided into three categories: (1) free movement in space without physical contact with surrounding objects, (2) contact that includes unbalanced reaction forces such as pressing on an object with the tip of a finger and (3) contact that includes balanced internal forces such as holding an object between the thumb and index finger [23, 24].

Alternatively, classification of haptic interactions can be based on whether the user perceives and manipulates objects directly or using a tool. The complexity of haptic displays highly depends on type of interactions to be simulated by the interface. An ideal haptic display designed for realistic simulation would have to be capable of simulating the handling of various tools. Such a display would measure limb position and display reaction forces. It would have a unique shape (e.g. exoskeleton) that could be used for different applications by adapting the device controller. However, complexity of human limbs and exceptional sensitivity of skin receptors together with inertia and friction of the device mechanism and constraints related to sensing and actuation of the display prevent the implementation of such complex devices based on the state-of-the-art technology.

Haptic displays can be divided into grounded (non-mobile) devices and mobile devices. Haptic perception and manipulation of objects require application of force vectors on the user at different points of contact with an object. Consequently, equal and opposite reaction forces act on the haptic display. If these forces are internally balanced, as while grasping an object with the index and thumb fingers, then mechanical grounding of the haptic display against the environment is not required. In the case of internally unbalanced forces, as while touching an object with a single finger, the haptic display must be grounded to balance the reaction forces. This means that a haptic display placed on a table is considered a grounded device while an exoskeleton attached to the forearm is a mobile device. If the exoskeleton is used for simulating contact with a virtual object using a single finger, forces that would in the real world be transferred across the entire human musculoskeletal system are now transferred only to the forearm.

Figure 7.21 shows classification of haptic displays based on their workspace, power and accuracy.

The use of grounded haptic displays has several advantages while executing tasks in a virtual environment. Such displays can render forces that originate from grounded sources without distortions and ambiguities. They may be used for displaying geometric properties of objects such as size, shape and texture as well as dynamic properties such as mass, stiffness and friction.

The main advantage of mobile haptic displays is their mobility and therefore, larger workspace. In order to illustrate ambiguities while displaying reaction forces using a mobile haptic display, two examples are analyzed in Fig. 7.22: grasping a virtual ball and pressing a virtual button. In the case of a virtual ball grasped with the thumb and index fingers, forces acting on the tip of fingers are all that is necessary for a realistic presentation of size, shape and stiffness of a virtual object.

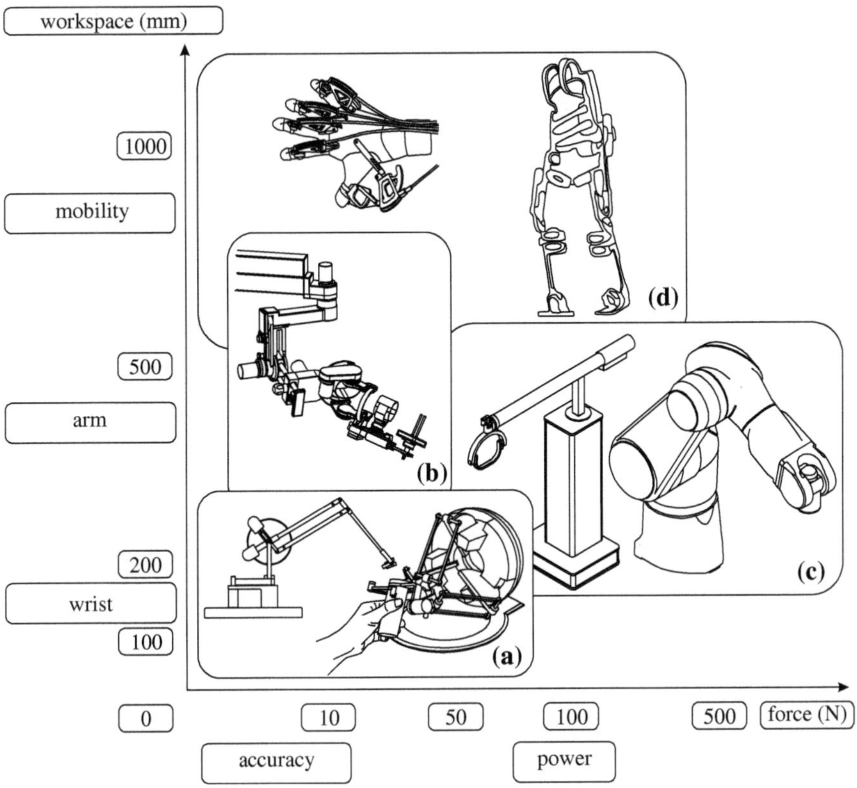

Fig. 7.21 Haptic displays classified based on their workspace, power and accuracy: **a** haptic displays for hand and wrist, **b** arm exoskeletons, **c** haptic displays based on industrial manipulators **d** mobile haptic displays

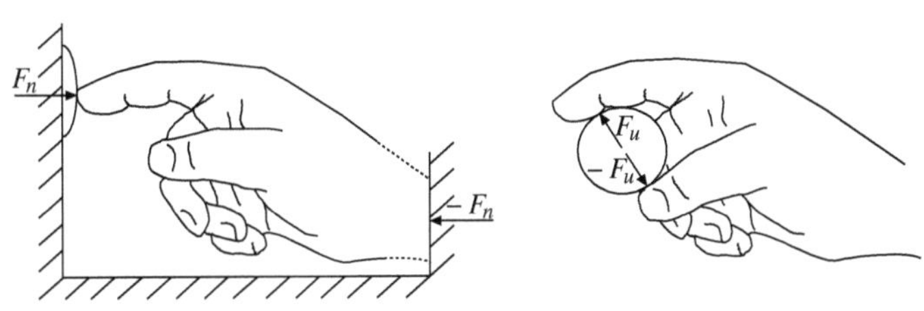

Fig. 7.22 Internally balanced forces F_u when holding a ball and unbalanced forces F_n when pressing a button

Only internally balanced forces act between the fingers and the ball. On the other hand, when pressing against a button, the user does not only feel the forces acting on the finger. The reaction forces also prevent further hand movement in the direction of the button. In this case the ungrounded haptic display can simulate the impression of a contact between the finger and the button, but it cannot generate the reaction force that would stop the arm movement [25].

Haptic displays may exist in the form of desktop devices, exoskeleton robots or large systems, which can move high loads. Given the diversity of haptic feedback (tactile, proprioceptive, thermal) and the different parts of the body to which the display can be coupled, displays have diverse properties. Properties of haptic displays determine the quality of the virtual reality experience. Design of haptic displays requires compromises which ultimately determine the realism of virtual objects. Realism defines how realistically certain object properties (stiffness, texture) can be displayed compared to direct contact with a real object. Low refresh rate of a haptic interface, for example, significantly deteriorates the impression of simulated objects. Objects generally feel softer and contact with objects results in annoying vibrations which affect the feeling of immersion. A long delay between an event in a virtual environment and responses of the haptic display furthermore degrades the feeling of immersion. Since haptic interactions usually require hand-eye coordination, it is necessary to reduce both visual as well as haptic latencies and synchronize both displays.

Kinaesthetic cues represent a combination of sensory signals that enable human awareness about joint angles as well as muscle length and tension in tendons. They enable the brain to perceive body posture and the environment around us. The human body consists of a large number of joints and segments that all have receptors that provide kinaesthetic information. Therefore, it becomes impossible to cover all possible points of contact with the body with a single haptic display.

Tactile cues originate from receptors in the skin that gather information resulting from local contact. Mechanoreceptors provide accurate information about the shape and surface texture of objects. Thermoreceptors perceive heat flow between the object and the skin. Electroreceptors perceive electrical currents that flow through the skin and pain receptors perceive pain due to skin deformation or damage.

Grounding of force displays provides support against forces applied by the user. A display grounded relative to the environment restricts the movement of the user to a space of absolute positions and orientations. Such display restricts the movement between the user and the outside environment. If a display is attached only to the body of the user, it is limited in its ability to render forces that originate from grounded sources in the environment. Such display can only render forces that are internally balanced between the device and the user.

The user's mobility is restricted with the use of haptic displays that are grounded to the environment. On the other hand, mobile displays allow users to move freely in a large space.

The number of haptic channels is usually very limited as a result of mechanical complexity. However, combinations of haptic displays can, for example, be used to enable bimanual manipulation.

The number of degrees of freedom and their characteristics determine the workspace of a haptic interface. To reach any pose in a three-dimensional space, a display with six degrees of freedom is required. Displays with less than four degrees of freedom are usually limited to rendering position and force, not orientation and torque.

Physical form is determined by the part of the haptic display with which the user interacts. The form of the haptic display can be a control prop, which represents a simple shape (stick, ball), a control prop in the form of an object (a pen, tweezers), or an amorphous form which varies depending on the needs of the display (gloves).

Spatial and temporal resolution determine the quality of haptic interaction. The ability of a human sensory system to distinguish between two different nearby tactile stimuli varies for different parts of the body. This information defines the required spatial resolution of a haptic display. Temporal resolution is defined by the refresh rate of a haptic display control system. A low refresh rate usually causes virtual objects to feel softer and collisions with virtual objects often result in annoying vibrations.

Safety is of utmost importance in dealing with haptic displays in the form of robots. High forces, which may be generated by haptic devices, can damage the user in case of a system malfunction.

References

1. Minsky M, Ouh-Young M, Steele OFB, Behensky M (1990) Feeling and seeing: issues in force display. Comput Graphics 24:235–443 (ACM Press)
2. Barfield W, Furness TA (1995) Virtual environments and advanced interface design. Oxford University Press, New York
3. Duke D, Puerta A (1999) Design. Specifications and verification of interactive systems. Springer, Wien
4. Mihelj M, Podobnik J (2012) Haptics for Virtual Reality and Teleoperation. Springer
5. Jones LA (2000) Kinesthetic sensing. Human and machine haptics. MIT Press, Cambridge
6. Biggs SJ, Srinivasan MA (2002) Handbook of virtual environments, chap haptic interfaces. LA Earlbaum, New York
7. Lederman SJ, Klatzky R (2009) Haptic perception: a tutorial. Attention Percept Psychophysics 71:1439–1459
8. Barraff D (1994) Fast contact force computation for nonpenetrating rigid bodies. Computer Graphics Proceedings, SIGGRAPH, Orlando, pp 23–34
9. Gottschalk S (1997) Collision detection techniques for 3D models. Cps 243 term paper, University of North Carolina
10. Lin M, Gottschalk S (1998) Collision detection between geometric models: a survey. In: Proceedings of IMA conference on mathematics on surfaces, pp 11–19
11. Adachi Y, Kumano T, Ogino K (1995) Intermediate representation for stiff virtual objects. In: Proceedings of virtual reality annual international symposium, pp 203–210
12. Konig H, Strohotte T (2002) Fast collision detection for haptic displays using polygonal models. In: Proceedings of the conference on simulation and visualization, Ghent, pp 289–300
13. Okamura AM, Smaby N, Cutkosky MR (2000) An overview of dexterous manipulation. In: Proceedings of the IEEE international conference on robotics and automation, pp 255–262
14. Salisbury JK, Brock D, Massie T, Swarup N, Zilles C (1995) Haptic rendering: programming touch interaction with virtual objects. Symposium on interactive 3D graphics, Monterey, USA, pp 123–130
15. Basdogan C, Srinivasan MA (2001) Handbook of virtual environments: design, implementation, and applications, chap. haptic rendering in virtual environments, Lawrence Erlbaum Associates, New Jersey, pp 117–134
16. Kazerooni H, Her MG (1994) The dynamics and control of a haptic interface device. IEEE Trans Rob Autom 20:453–464

17. Hogan N (1989) Controlling impedance at the man/machine interface. In: Proceedings of the IEEE international conference on robotics and automation, pp 1626–1631
18. Carignan CR, Cleary KR (2000) Closed-loop force control for haptic simulation of virtual environments. Haptics-e 1(2):1–14
19. Hannaford B, Venema S (1995) Virtual environments and advanced interface design, chap. Kinesthetic displays for remote and virtual environments, Oxford University Press Inc., New York, pp 415–436
20. Youngblut C, Johnson RE, Nash SH, Wienclaw RA, Will CA (1996) Review of virtual environment interface technology. Ida paper p-3786, Institute for Defense Analysis, Virginia, USA
21. Burdea G (1996) Force and touch feedback for virtual reality. Wiley, New York
22. Hollerbach JM (2000) Some current issues in haptics research. In: Proceedings of the IEEE international conference on robotics and automation, pp 757–762
23. Bar-Cohen Y (1999) Topics on nondestructive evaluation series, vol 4: automation, miniature robotics and sensors for non-destructive testing and, evaluation, The American Society for Nondestructive Testing, Inc
24. Hayward V, Astley OR (1996) Performance measures for haptic interfaces. Robotics Research, pp 195–207
25. Richard C, Okamura A, Cutkosky MC (1997) Getting a feel for dynamics: using haptic interface kits for teaching dynamics and control. In: Proceedings of the ASME IMECE 6th annual symposium on haptic interfaces, Dallas, TX, USA, pp 15–25

Chapter 8
Augmented Reality

Abstract Augmented reality is defined as augmenting an image of the real world with a computer-generated image that provides additional information. After introducing the basic definitions, the chapter covers the challenge of modeling the real environment. This process generally has two phases: sensing the information from the environment using various sensors, then reconstructing the environment. The chapter also describes the specific displays and user interfaces used in augmented reality. It concludes with a few representative applications, from medicine to advertising.

8.1 Definition

The goal of virtual reality is to replace sensations from the real world with artificial sensations that originate from a virtual world. In an ideal virtual reality system, the human is thus completely immersed into the virtual world and does not perceive the real world at all. However, no-one says that both worlds can't be presented to the user at the same time: some information from the real world and some from the virtual world. The virtual environment thus doesn't envelop the user completely, allowing him/her to maintain a feeling of presence in the real world.

In 1994, Milgram and Kishino [1] introduced the *reality-virtuality continuum* to describe such *mixed* realities . The continuum defines different mixtures of real and virtual worlds (Fig. 8.1). Between the purely real and virtual environments, we can thus also find augmented reality (real world with additional virtual information) and augmented virtuality (virtual world with additional real information). Today, augmented reality is much more prevalent than augmented virtuality and already has many important applications.

Augmented reality is defined as augmenting an image of the real world (seen by the user) with a computer-generated image that enhances the real image with additional information. Besides combining the real and virtual worlds, an augmented reality system must also allow interaction in real time and track both real and virtual objects

M. Mihelj et al., *Virtual Reality Technology and Applications*,
Intelligent Systems, Control and Automation: Science and Engineering 68,
DOI: 10.1007/978-94-007-6910-6_8,

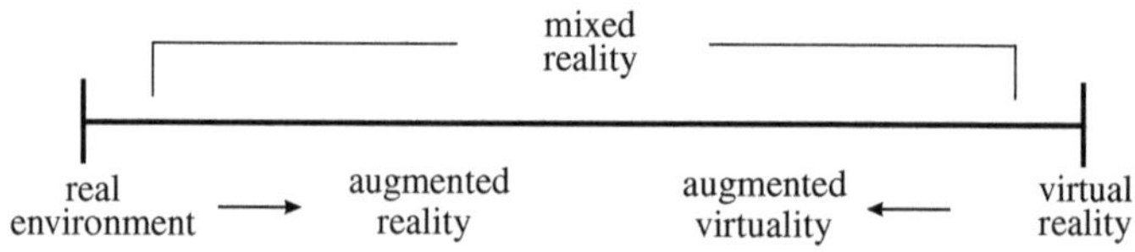

Fig. 8.1 Reality-virtuality continuum [1]

in three-dimensional space [2]. Augmented reality thus faces additional challenges compared to virtual reality. Though it needs to show fewer virtual objects, it must regularly refresh them as well as synchronize the real and virtual images with minimal delays.

The purpose of augmented reality is to improve user perception and increase his/her effectiveness through additional information. The user retains awareness of the real world, but in an ideal augmented reality would not be able to tell the difference between information from the real and from the virtual world. If the information is to be successfully combined, virtual objects must act in a physically appropriate manner. If a real and a virtual object collide, both must react appropriately. Furthermore, virtual objects must block the view of real objects and cast shadows on them. All this can only be achieved with an accurate model of both the real and the virtual environment.

8.2 Modeling the Real Environment

Modeling the real environment usually has two phases: first sensing the information from the environment, then reconstructing the environment.

8.2.1 Sensing the Environment

Information from the real environment can be obtained using different sensing technologies: digital cameras, accelerometers, global positioning systems (GPS), ultrasonic sensors, magnetometers, lasers, radio waves etc. Compared to sensors for virtual reality, sensors for augmented reality require a higher accuracy and greater range since they may also be used e.g. outdoors. Of course, sensing is much easier indoors since outdoor sensors need to be more mobile and resistant to damage. Furthermore, buildings can easily be modeled in advance, and their lighting or temperature can be controlled.

Sensors in augmented reality are divided into active and passive ones. With passive sensors, no equipment needs to be mounted on the object we wish to detect; everything is done by the sensor. Such systems are more user-friendly since objects don't need to be additionally equipped with cumbersome cables, but accurate passive sensing requires expensive equipment and complex software. Active sensors involve a device (such as a marker) placed on the object we wish to track. This makes tracking easier, but the devices need to be placed on all objects.

Popular sensor systems in augmented reality include [3]:

- *Cameras with passive tracking* are normal videocameras that record images of the environment, then use image analysis methods (edge search, comparison to previously recorded images) to extract objects and determine their position in the environment. A training phase is usually necessary for successful object recognition. It involves showing different objects to the camera from different angles, thus allowing it to recognize them later.
- *Cameras with active tracking* also record images of the environment, but they do not try to recognize objects in the image. Instead, they only search for special (*markers*) that were previously placed on objects. These markers either have a special shape or emit light (visible or infrared), so they can be easily recognized by computers. Cameras with active tracking are more accurate than those with passive tracking, but require the markers to be placed in advance.
- *Ultrasonic sensors* detect objects using ultrasonic waves (usually 40 kHz) emitted into the environment. There are two possible implementations. In the first one, the ultrasound emitter is attached to the object while the receivers are arrayed around the room. The object's position can be calculated from the time it takes the ultrasound wave to reach the different receivers. In the second implementation, both the emitter on the object and a fixed emitter in the room emit ultrasound waves of the same frequency. These waves are measured using multiple receivers arrayed around the room. The receivers measure the sum of both waves, which is different depending on the position of both emitters (phase delay).
- *Inertial sensors* are a combination of accelerometers and gyroscopes attached to the object we wish to track. If the object's initial position is known, it can theoretically be tracked by integrating the measured acceleration. In practice, it is necessary to minimize measurement errors, as they are otherwise also integrated and thus result in inaccurate measurements. Inertial sensors are thus frequently combined with magnetometers, which measure the Earth's magnetic field and give a reference absolute orientation. Magnetometers themselves can provide certain information about object positions, but combining them with accelerometers and gyroscopes allows more accurate tracking.
- *Global positioning systems* calculate their position based on radio signals transmitted by a system of artificial satellites orbiting the Earth. Each satellite transmits information about its own position, the positions of the other satellites, and the time at which the signal is emitted. A receiver needs a connection with at least four satellites to calculate its position. Receivers need to be attached to all objects we wish to track, and the tracking quality depends on the sensitivity and accuracy of the receiver as well as the quality of the connection with the satellites. This quality is poor inside buildings or near very high buildings, among other places.
- *Hybrid systems* combine multiple types of sensors and thus compensate for the disadvantages of each individual type. Global positioning systems can, for example,

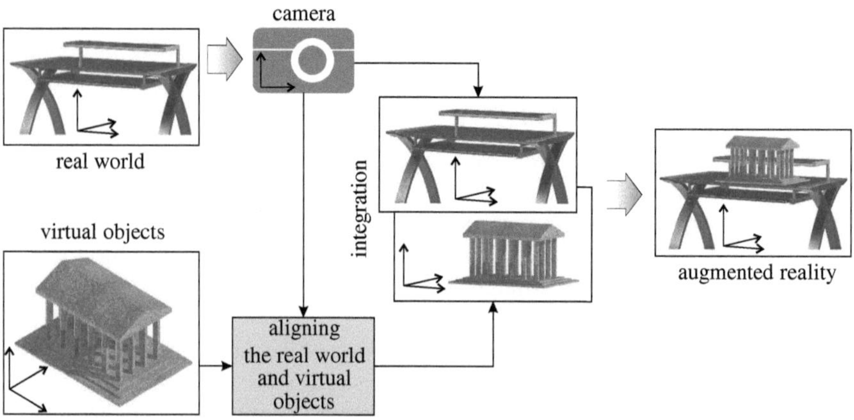

Fig. 8.2 Integration of the real and virtual environments. The different coordinate systems need to be properly aligned

be combined with inertial systems that temporarily track the object's motion when no connection to a satellite is available. Similarly, inertial sensors can be combined with cameras since cameras perform better with slow motions while inertial sensors perform better with fast motions.

8.2.2 Environment Reconstruction

Once the positions of the user, display and objects in the real environment are known, it is possible to create a three-dimensional model of the real environment and integrate it with a model of the virtual environment (Fig. 8.2). The integrated model then allows e.g. collisions between real and virtual objects to be calculated. The mathematical tools needed to reconstruct the real environment are identical to the previously described methods for calculating interactions between objects in virtual reality (Chap. 3), and will thus not be separately described here.

8.3 Displays

8.3.1 Visual Displays

Similarly to virtual reality, the visual appearance is probably the most important component of augmented reality. The basic visual display technologies are similar to

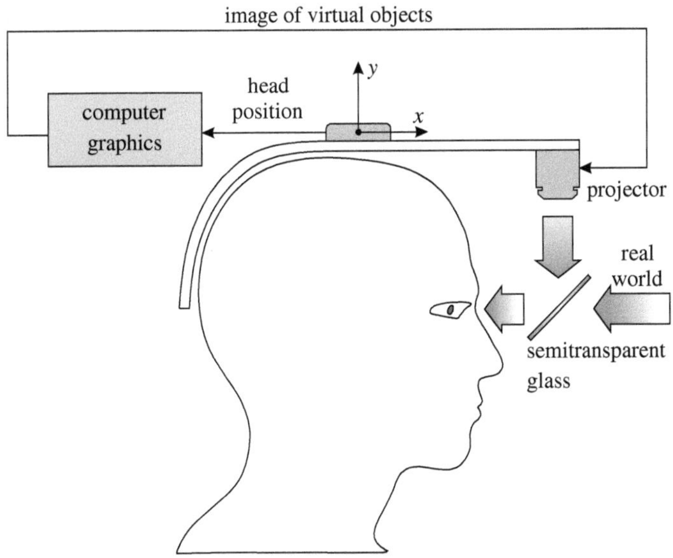

Fig. 8.3 Direct (see-through) head-mounted display

those in virtual reality, but there are additional challenges due to the need for mobility and the integration of information from both real and virtual environments.

8.3.1.1 Head-Mounted Displays

Head-mounted displays (described in detail in Chap. 5) consist of glasses with one or two screens that display the virtual image. The left and right eyes see slightly different images, creating an illusion of depth. Augmented reality presents the additional challenge of having to combine images of the real and virtual environments. This challenge can be solved in two ways.

Direct displays use semitransparent glass as the screens (Fig. 8.3). The user can thus see the real world through the screens, making the displays simple and natural. A special case of such displays are retinal displays, which project an image of the virtual environment directly on the retina. They offer a higher-quality image than classic direct displays, but are not yet widespread.

Indirect displays use opaque screens and do not offer the user a direct view of the real world (Fig. 8.4). Instead, the image of the real world is captured by a camera mounted on the display. This image is transferred to a computer and combined with the image of the virtual world, and the integrated image is then shown on the screen. Such an implementation is more complex, but offers more flexible integration of real and virtual information. It is, for instance, possible to temporally synchronize the images and thus avoid problematic delays.

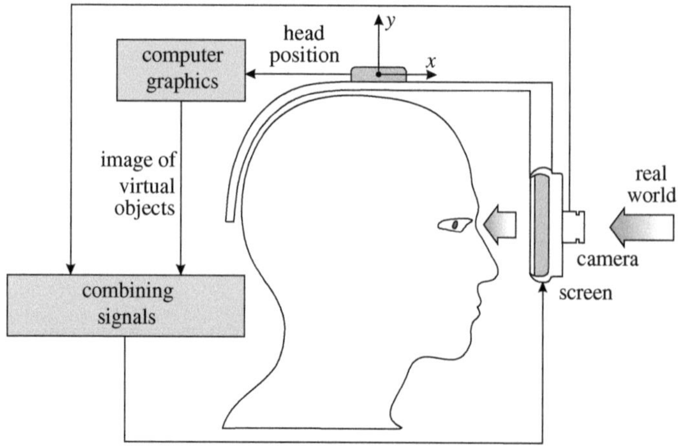

Fig. 8.4 Indirect head-mounted display

8.3.1.2 Handheld Displays

Handheld displays are built into small, portable device such as smartphones or tablet computers. A camera is usually built into the other side and captures an image of the real world. By displaying this image on the screen, it gives the user the impression of looking *through* the device. These displays usually also include accelerometers, digital compasses or global positioning systems, making them very mobile and suitable for outdoor use. However, due to their small screen they usually allow only a two-dimensional virtual image and do not create a feeling of virtual presence.

8.3.1.3 Spatial Displays

Spatial displays create the virtual component of augmented reality on the surface of objects in the environment. This is usually done with projectors or holograms that can either be limited to a single object (e.g. table of wall) or cover the entire room with augmented reality. In both cases, a model of the room and the objects in it is required for accurate projection. Spatial displays offer both two-dimensional and three-dimensional images, and can also be used by several people simultaneously.

8.3.2 Sound Displays

Sound displays in augmented reality are mostly limited to the kind of headphones and speakers seen in normal virtual reality. However, some displays also incorporate so-called *haptic sound*: sound felt through vibrations. This is generally used in headphones and mobile devices in order to increase realism and augment user interfaces.

8.3.3 Other Displays

In principle, augmented reality can stimulate all five senses, but most practical systems focus on sight and hearing. Haptic feedback appears mainly as part of user interfaces while smell and taste are rarely seen in both virtual and augmented reality. However, some examples do exist. The most noteworthy are food simulators, which are equipped with scented, tasty fluids. They offer the user a normal piece of food sprayed with one of the fluids, giving the user the impression of eating a different type of food in augmented reality than in the real world.

8.4 User Interfaces

Just like virtual reality, augmented reality must offer the user the possibility of interacting with virtual objects. Typical user interfaces include [4, 5]:

- *Tangible interfaces* allow interaction with the virtual world via physical objects and tools: pointers, gloves, pens, haptic robots etc.
- *Collaborative interfaces* use multiple displays and interfaces, allowing several users to work together. These users can all be in the same place or at any distance from each other.
- *Hybrid interfaces* combine multiple complementary interfaces and thus allow many interaction options. Since they are very flexible, they are suitable for spontaneous environments where we do not know how the user will wish to communicate with the augmented reality system.
- *Multimodal interfaces* combine tangible interfaces with natural forms of interaction such as speech, arm movements and gaze.

8.5 Applications

8.5.1 Entertainment and Education

Augmented reality is an excellent opportunity for games that augment the real game (e.g. a board) with sounds and visual stimuli. These can make the game more interesting or even help the player by, for example, giving a warning when the desired move is invalid. A simple example is chess with virtual figures that can be moved with a pointer. A similar principle is used by videogames that use various interfaces to combine information from the real and virtual environments. Some games even allow collaboration between many people using multiple displays and interfaces. Perhaps the first example of such a game was the British television show Knightmare, where a group of children must accomplish a certain goal in augmented reality. One of the children travels through a virtual world with real opponents (players) while the other children observe on screens and give instructions.

Augmented reality games can also be used for educational purposes. The US army, for instance, allows soldiers to train with real weapons and virtual opponents that react to the soldier's movements. More peaceful games may ask the player to solve various physical or mental challenges, thus teaching certain skills.

The user of an augmented reality system is not necessarily an active participant in the game; he or she may only be a passive observer that obtains additional information via augmented reality. The concept is often seen in sports broadcasts where the video from the playing field (real information) is combined with displays of the current score, statistical data about the players and so on. If the broadcast is shown on a computer, the viewer may be able to select individual players and viewpoints, thus obtaining the most desired information. The same concept can also easily be used for education: at museums and other sights, augmented reality can offer additional information about the user's location and the objects seen there.

8.5.2 Medicine

Augmented reality has been extensively used to train doctors similarly to educational games from the previous subsection. Furthermore, it is a valuable tool even for experienced doctors since it can offer additional information in critical situations. For instance, during surgery the computer can project an image of the patient's internal organs on the surface of the skin and thus help determine the exact location of an incision. During diagnostic procedures, the computer can also project internal organs onto the skin, thus letting the doctor better examine critical spots and estimate the patient's health.

8.5.3 Design and Repairs

Just like a surgeon can obtain information about the patient's internal organs during surgery, an engineer or repairman can obtain information about a machine's internal parts while assembling or repairing it. Here, augmented reality projects a blueprint or other information (e.g. temperature of individual parts) directly onto the device, thus allowing easier analysis of individual parts and a better overview of the device as a whole.

Augmented reality can also be used to design complex machines. Actual components can be combined with virtual components that we wish to test. We can thus

Fig. 8.5 The QR-code is a very simple example of augmented reality in marketing

also quickly determine whether a component is suitable, whether the model of the machine accurately corresponds to the real machine, how the completed product would look and so on.

8.5.4 Navigation

When we're traveling, augmented reality can help us reach our goal by providing us with additional information. If traveling by foot, we can photograph the road with a handheld device. The augmented reality system then finds known landmarks on the image and uses them to determine the best route to take. If traveling by car, augmented reality can be projected directly onto the windshield and provide the driver with information such as road and weather conditions.

8.5.5 Advertising

Augmented reality was first used for advertising in the automotive industry. Some companies printed special flyers that were automatically recognized by webcams, causing a three-dimensional model of the advertised car to be shown on the screen. This approach then spread to various marketing niches, from computer games and movies to shoes and furniture. The ubiquitous QR-code (Fig. 8.5) is a very simple example of such augmented reality: a black-and-white illustration that turns into more complex information when analyzed by a mobile phone or computer.

An example of more complex augmented reality is virtually trying on shoes. The user wears a special pair of socks, then walks in front of a camera and sees his/her own image on the screen wearing a desired pair of shoes. The model, color and accessories of the shoes can be changed in an instant, allowing the user to easily find the most attractive footwear.

References

1. Milgram P, Kishino AF (1994) A taxonomy of mixed reality visual displays. IEICE Trans Inf Syst E77–D(12):1321–1329
2. Azuma R (1997) A survey of augmented reality. Presence Teleoperators Virtual Environ 6: 355–385
3. Costanza E, Kunz A, Fjeld M (2009) Mixed reality: a survey. Lecture notes on computer science, vol 1. Springer
4. Carmigniani J, Furht B, Anisetti M, Ceravolo P, Damiani E, Ivkovic M (2011) Augmented reality technologies, systems and applications. Multimedia Tools Appl 51:341–377
5. van Krevelen DWF, Poelman R (2010) A survey of augmented reality technologies, applications and limitations. Int J Virtual Reality 9:1–20

Chapter 9
Interaction with a Virtual Environment

Abstract Interaction with a virtual environment is the most important feature of virtual reality, which requires the computer to respond to user's actions. The mode of interaction with a computer is determined by the type of the user interface. The three most important aspects of interaction are manipulation of objects within the virtual environment, navigation through the virtual environment and communication with other entities existing in the virtual environment. All three aspects are the subject of this chapter.

Interaction with a virtual environment is the most important feature of virtual reality. Interaction with a computer-generated environment requires the computer to respond to user's actions. The mode of interaction with a computer is determined by the type of the user interface. Proper design of the user interface is of utmost importance since it must guarantee the most natural interaction possible. The concept of an ideal user interface uses interactions from the real environment as metaphors through which the user communicates with the virtual environment.

Interaction with a virtual environment can be roughly divided into *manipulation*, *navigation* and *communication*. Manipulation allows the user to modify the virtual environment and to manipulate objects within it. Navigation allows the user to move through the virtual environment. Communication can take place between different users or between users and intermediaries in a virtual environment.

9.1 Manipulation Within Virtual Environment

One of the advantages of operating in an interactive virtual environment is the ability to interact with objects or manipulate objects in this environment [1]. The ability to experiment in a new environment, real or virtual, enables the user to gain knowledge about the functioning of the environment. Some manipulation methods are shown in Fig. 9.1.

M. Mihelj et al., *Virtual Reality Technology and Applications*, Intelligent Systems, Control and Automation: Science and Engineering 68, DOI: 10.1007/978-94-007-6910-6_9, © Springer Science+Business Media Dordrecht 2014

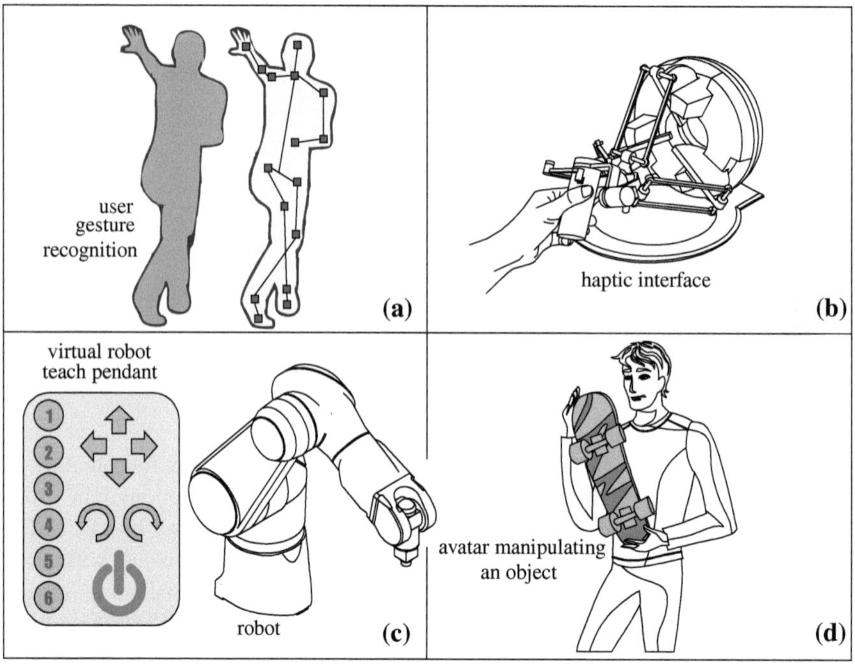

Fig. 9.1 Manipulation methods: **a** direct user control (gesture recognition), **b** physical control (buttons, switches, haptic robots), **c** virtual control (computer-simulated control devices) and **d** manipulation via intelligent virtual agents

Direct user control (Fig. 9.1a) allows a user to interactively manipulate an object in a virtual environment the same way as he would in the real environment. Gestures or gaze direction enable selection and manipulation of objects in virtual environment.

Physical control (Fig. 9.1b) enables manipulation of objects in a virtual environment with devices from real environment (buttons, switches, haptic robots). Physical control allows passive or active haptic feedback.

Virtual control allows manipulation of objects through computer-simulated devices (simulation of real-world devices—virtual buttons, steering wheel—Fig. 9.1c) or avatars (intelligent virtual agents—Fig. 9.1d). The user activates a virtual device via an interface (real device), or sends commands to an avatar that performs the required action (voice commands or through the use of gestures). The advantage of virtual control is that one real device (for example a haptic robot) activates several virtual devices.

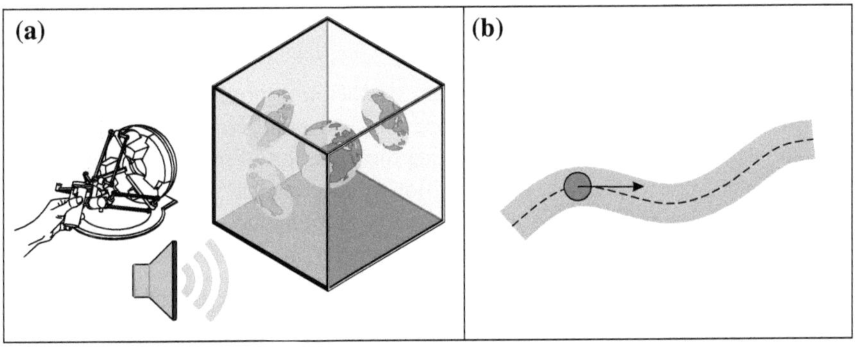

Fig. 9.2 Multimodal feedback (*left image*); virtual fixture that constrains movement of a ball along the tunnel (*right image*)

9.1.1 Manipulation Characteristics

Manipulation in a virtual environment is usually based on similar operations as in the real environment; however, interfaces and methods exist that are specific.

Feedback may exist in visual, aural or haptic form (Fig. 9.2a). However, one type of information can often be substituted with another (for example, haptic information can be substituted with audio cues). On the other hand, *virtual fixtures*, for example, allow easier and safer execution of tasks in a virtual environment. Virtual fixtures are software-generated movement constraints (Fig. 9.2b). As the name implies, virtual fixtures are not mechanical constraints, but are instead part of the control algorithms. The main purpose of the virtual fixtures is to guide or to confine movements commanded by the human operator. Virtual fixtures are used in applications where a task requires both precision and accuracy as well as the decision-making power and intelligence of the human operator [2].

The viewpoint defines the perspective from which the virtual environment is presented to the user. Selection of a viewpoint significantly affects the ability of manipulation and navigation in a virtual environment (Fig. 9.3). A first-person view refers to a graphical perspective rendered from the viewpoint of the player avatar (often used for presentation of the environment from the cockpit of a vehicle). It enables the user to look around the virtual environment. On the other hand, third-person refers to a graphical perspective rendered from a fixed distance behind and slightly above the player character. This viewpoint allows a user to see a more strongly characterized avatar.

9.1.2 Selection of Objects

Selection means definition of the direction of operation or choice of objects to be manipulated. Selection methods are based on hand gestures, gaze direction, head or torso orientation (primarily to determine the direction of movement), the use of input devices (such as joysticks) or definition of coordinates. Selection of objects is based on touch (an avatar or a virtual tool representation touches an object), gestures (indicate the object that is to be selected), a 3D cursor, menus, or speech recognition.

9.1.3 Manipulation of Objects

Manipulation of object shape and location is an important aspect of interaction. *Application of force to a virtual object* allows interactions such as object grasping, pushing, squeezing and hitting. A haptic robot allows a realistic presentation of forces acting on the object being manipulated. *Modification of the state of virtual control interfaces* allows the position of switches, buttons, sliders and other control functions implemented in a virtual environment to be changed. *Modification of object properties* allows quantities such as transparency, color, mass, density of objects to be changed. These are operations that do not replicate real-environment actions.

9.2 Navigation Within the Virtual Environment

Navigation represents movement in space from one point to another. It includes two important components: (1) *travel* (how the user moves through space and time) and (2) *path planning* (methods for determination and maintenance of awareness of position in space and time, as well as trajectory planning through space to the desired location).

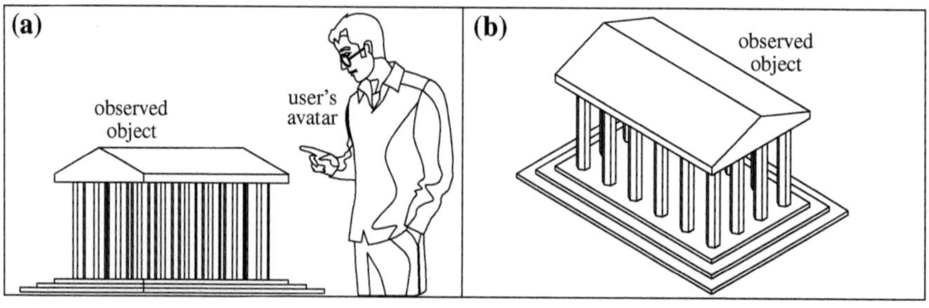

Fig. 9.3 What an external viewer would see (*left image*); what the avatar would see—first-person perspective (*right image*)

9.2.1 Path Planning

In determining the current position and path through space, it is important to generate a mental model of the virtual environment through which the user moves. Knowing the location and neighborhood is defined as position awareness. Creation of a mental model is based on different strategies that can be summarized as (1) *divide and conquer*—the virtual environment is divided into smaller subregions; the user learns features of each subregion and determines paths between subregions, (2) *global network*—is based on the use of landmarks that the user remembers; navigation is based in relation to these known landmarks, and (3) *gradual expansion*—is based on gradual memorization of the map of an entire area (the user starts with a small region that is gradually expanded outwards). Path planning may be assisted by maps, instrumental navigation, virtual fixtures or other features. A birds-eye view of the scene significantly simplifies navigation as well.

9.2.2 Traveling

In a virtual environment, where the area of interest extends beyond direct virtual reach of the user, traveling is one possibility for space exploration. Some traveling methods are shown in Fig. 9.4.

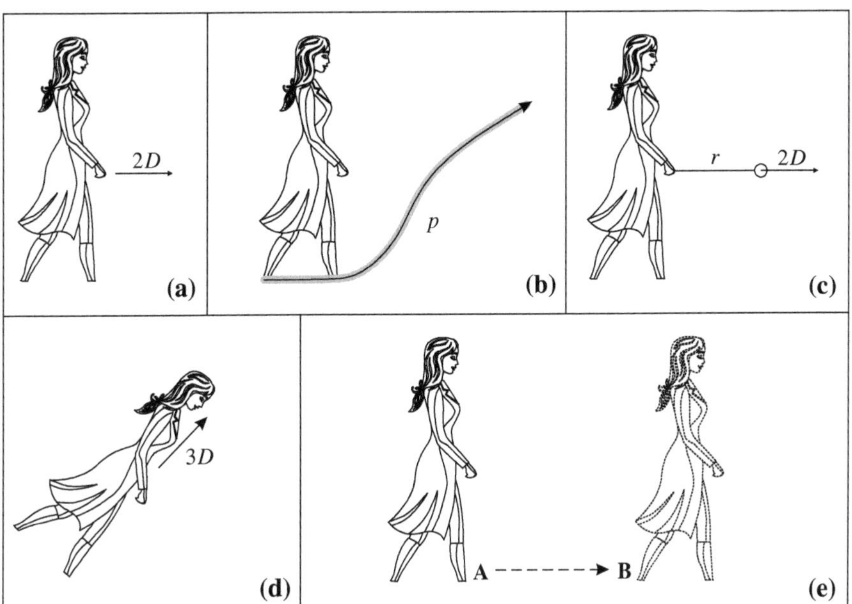

Fig. 9.4 Traveling methods: **a** locomotion, **b** path tracking, **c** towrope, **d** flying and **e** displacement

Physical locomotion is the simplest way to travel. It requires only tracking the user's body movement and adequate rendering of the virtual environment. The ability to move in real space also enables proprioceptive feedback, which helps to create a sense of relationships between objects in space. A device that tracks user movement must have a sufficiently large working area. *Path tracking* (virtual tunnel) allows the user to follow a predefined path in a virtual environment. The user is able to look around, but cannot leave the path. *The towrope method* is less constraining for the user than path tracking. The user is towed through space and may move around the coupling entity in a limited area. *Flying* does not constrain the user movement to a surface. It allows free movement in three-dimensional space. At the same time it enables a different perspective of the virtual environment. The fastest way of moving through a virtual environment is a simple *displacement* that enables movement between two points without navigation (the new location is reached instantly).

9.3 Interaction with Other Users

Simultaneous activity of more users in a virtual environment is an important property of virtual reality. Users' actions in virtual reality can be performed in different ways. If users work together in order to solve common problems, the interaction results in *cooperation*. However, users may also compete among themselves or interact in other ways. In an environment where many users operate at the same time, different issues need to be taken into account. It is necessary to specify how interaction between persons will take place, who will have control over the manipulation or communication, how to maintain the integrity of the environment and how the users communicate. Communication is usually limited to visual and audio modalities. However, it can also be augmented with haptics.

9.3.1 Shared Experience

The aim of interaction with other users is the exchange of information and experiences. Different methods exist for exchanging experiences with other users. Technologically, shared experience can be divided into two categories: all users are virtually present in a virtual environment or some users are outside observers and represent an audience that watches other users who are present in a virtual environment.

9.3.2 Cooperation with Interaction

Certain professions require cooperation between experts to accomplish a task within a specified time. In addition to manipulation tasks, where the need for physical power requires the cooperation of several persons, there are many tasks that require cooperation between experts such as architects, researchers or medical specialists. The degree of participation in a virtual environment may extend from zero, where users merely coexist in a virtual environment, to the use of special tools that allow users to simultaneously work on the same problem.

Interactive cooperation requires environmental coherency. This defines the extent to which the virtual environment is the same for all users. In a completely coherent environment, any user can see everything that other users do. It is often not necessary for all features of the virtual environment to be coherent. Coherency is of primary importance for simultaneous cooperation; for example, when more users work on a single object.

References

1. Bowman DA, Coquillart S, Froehlich B, Hirose M, Kitamura Y, Kiyokawa K, Stuerzlinger W (2008) 3d user interfaces: new directions and new perspectives. IEEE Comput Graph Appl 28:20–36
2. Abbott JJ, Marayong P, Okamura AM (2005) Haptic virtual fixtures for robot-assisted manipulation. In: 12th international symposium of robotics research (ISRR), pp 49–64

Chapter 10
Design of a Multimodal Virtual Environment

Abstract A functional virtual environment requires a number of components to be integrated in a single application. These include virtual environment model (geometry, dynamics, visual appearance, and acoustics), input devices enabling the user to interact with the virtual environment, as well as output modalities for providing the feedback to the user. In this chapter the basic concepts of virtual environment development are presented and the possible uses of such virtual reality applications are summarized.

In previous chapters we analyzed individual components of the virtual environment: geometry, dynamics, methods of visual, audio and haptic presentation and user interaction. A functional virtual environment requires all these elements to be integrated in a single application. In this section, we will therefore look at the basic concepts of virtual environment development and the possible uses of such applications.

The block scheme in Fig. 10.1 summarizes the core idea of the implementation and use of virtual reality applications. Virtual reality requires a three-dimensional model of the environment. Such a model can be developed specifically for the task. More importantly, however, three-dimensional models are often the result (byproduct) of development in a variety of human activities. For example, three-dimensional modeling is used in mechanical engineering to design mechanical components that constitute various devices, in architecture to design buildings and other structures, in medicine for diagnostic procedures and in geodesy for analysis and planning of the land use. All these three-dimensional models can provide a starting point for designing virtual environments.

To design a virtual environment, three-dimensional models should first be appropriately simplified in order to guarantee computation of the virtual environment and interactions within it in real-time. Then the designer should add various visual effects such as colors, lights and shadows. Since the virtual environment should have a dynamic component, it is necessary to introduce physics effects such as gravity, joints and other connections between objects as well as collision detection. In the virtual

M. Mihelj et al., *Virtual Reality Technology and Applications*,
Intelligent Systems, Control and Automation: Science and Engineering 68,
DOI: 10.1007/978-94-007-6910-6_10,

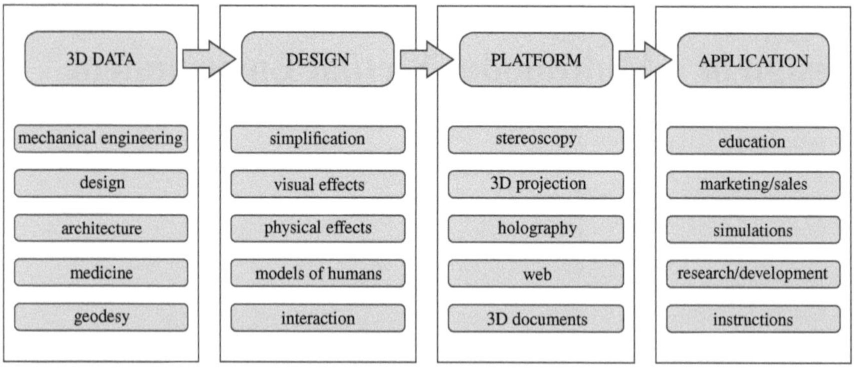

Fig. 10.1 A generalized concept for the use of virtual reality on different platforms and different applications

environment human models (avatars) can also be introduced, for example, to liven up the presentation of architecture of a building. The possibility of user interaction with the virtual environment is also important.

The virtual environment can be displayed on a variety of platforms, ranging from stereoscopic and holographic systems to web applications and interactive 3D documents.

Each virtual environment has its own purpose, its added value. Virtual environments can be efficiently used in education (presentation of systems through interactive simulations rather than using pictures in textbooks), they can be used for marketing and sales purposes (presentation of the product characteristics, product configuration based on the user's needs), they are important in areas of research and development (where the researcher can simulate and analyze events on the level of a model), but can also be used for interactive instructions for the use of various devices.

In the next sections we will analyze some examples of virtual environments and their use for various purposes.

10.1 Interactive Computer Game

The first presented scenario is a virtual game of table hockey. This virtual environment is relatively simple and consists of a hockey table, puck and two objects that represent the user's handle and the opponent (Fig. 10.2).

Different coordinate frames can be identified in any virtual environment: a coordinate frame of the virtual environment itself (a reference coordinate for all other entities), a coordinate frame that defines the viewpoint of the user (how the user perceives the virtual environment) and coordinate frames that define positions (and orientations) of objects (each object with its own coordinate frame) in the virtual

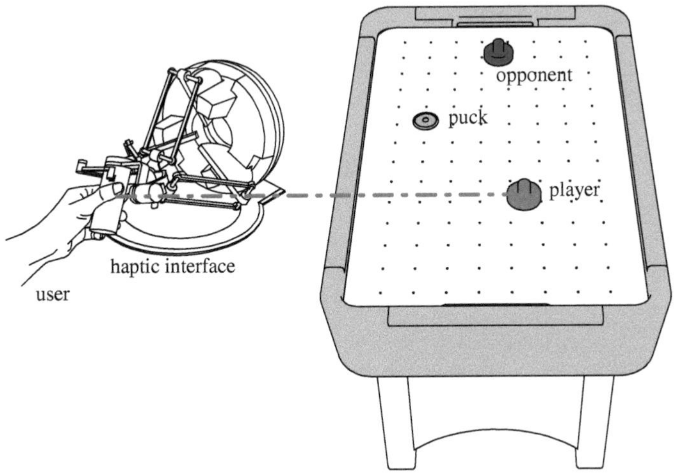

Fig. 10.2 User plays a game of table hockey through the use of a haptic interface. The user is coupled to the haptic device end-effector

environment. Movements of the user and objects are achieved by transformations of their respective coordinate frames.

In this case, we assume that the interaction between user and the virtual environment is implemented in the form of haptic interaction, visual and auditory feedback. The haptic robot end-effector is coupled to the object that represents the user's handle in the virtual environment. The haptic interface provides feedback to the human about forces occurring during collisions in the virtual environment, while the user controls his virtual handle through the movements of the haptic interface.

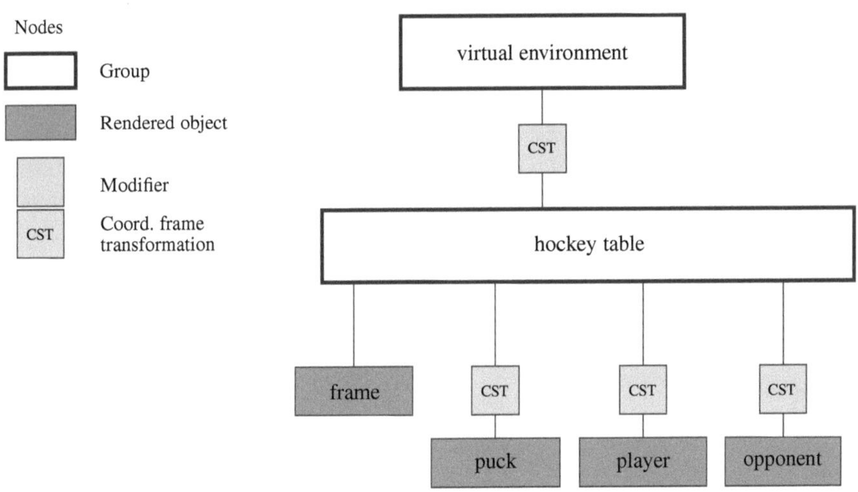

Fig. 10.3 Scene graph for table hockey

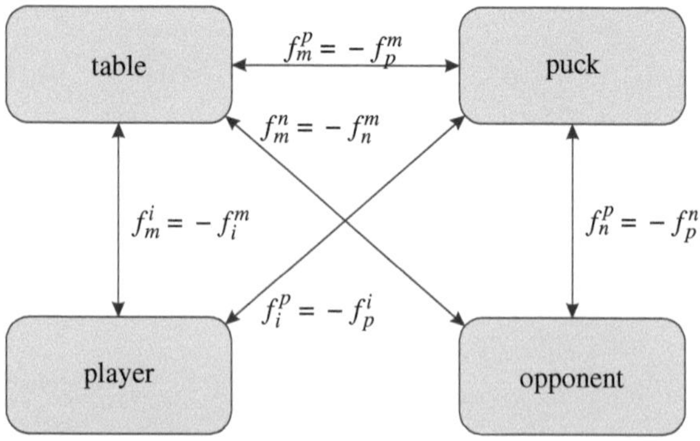

Fig. 10.4 Physical interactions between objects in a virtual environment and collision forces. The player and opponent cannot collide

Figure 10.3 shows the scene graph for a virtual game of table hockey. The scene graph is relatively simple due to the small number of virtual objects. A hockey table is placed in a virtual environment and playing objects (puck and two handles) are positioned on the table. Each individual element can move relative to the table, so blocks representing transformation of coordinate frames are placed between the *table* block and *playing object* blocks.

Objects in a virtual environment are defined by their properties, which determine their geometry, dynamics, visual and acoustic appearance: dimensions, weight, inertia, friction coefficient, stiffness, color and texture, sounds (during collisions).

Since there are three objects on the hockey table that can move independently of each other, it is necessary to detect collisions between them. Figure 10.4 shows a graph of possible collisions between objects. We assume that all collisions between objects are possible. Only the player and the opponent cannot collide because each is limited to its own half of the table. As a result of collisions reaction forces can be calculated. Forces always act on both colliding objects, but with the opposite sign. Forces acting on the table can be neglected since we assume that the table is grounded (an object with an infinite mass), which prevents its movement.

In addition to collision forces, it is necessary to consider also forces resulting from the interactions with the medium in which the virtual object is moving (air resistance, friction with the table) and the gravitational force field. All forces are then summed up and the resultant force is used to compute object movement based on object's dynamic properties.

Finally, computer games come to life through the use of 3D technologies. Stereoscopic displays with spatial effect put the player in the center of the action.

10.2 Simulated Operation of Complex Systems

As technology progresses, systems become more complex and their use requires many new skills. Virtual reality is a medium that is suitable for training operators of such systems. Thus, it is possible to practice flying an aircraft, controlling a crane, ship navigation, surgical procedures and many other tasks in a virtual environment. Such environments are also important in the field of robotics. Their advantage is not only that they enable simulation of a robotic cell. If a virtual environment contains software modules that behave as a real robot controller, it is possible to write and validate robot programs, which are only transferred to a real robot at a later stage. This saves time required for programming, since robot teaching is done in a simulation environment offline while the robot can still be used in a real environment in the mean time. At the same time, a virtual environment enables verification of software correctness before the program is finally transferred to the robot. Simulation-based programming may thus help avoid potential system malfunctions and consequent damage to the mechanism or robot cell.

Figure 10.5 shows a robotic cell consisting of a robot and an eccentric press. In this case, the robot acts as a device tending the press by inserting raw material and removing semifinished products. Shaded in gray are shown user interface devices (robot teach panel) for the robot and the press. Buttons on teach boxes simulate the

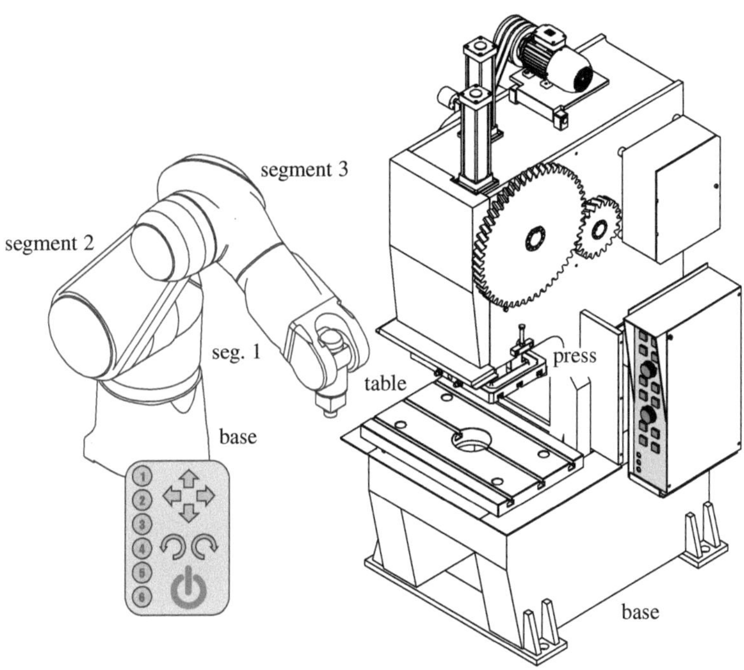

Fig. 10.5 Robot tending an eccentric press. Interfaces that are part of the virtual environment and enable interaction with the user (control of the robot and press) are marked in grey color

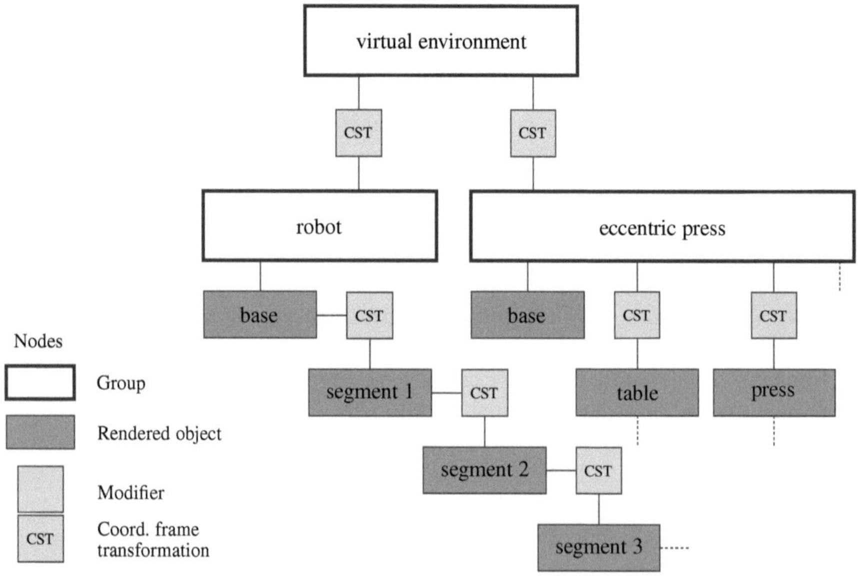

Fig. 10.6 Partial scene graph (most of the elements are omitted due to the model complexity) of a robot tending an eccentric press

operation of real controls. In principle, it is possible to also control the device through the physical form of a user interface (teach box), which is connected with a virtual environment. In such cases, interaction with the system becomes even more realistic. Through the teaching unit for the robot it is, for example, possible to program a robot as it would be done on the real system. A robot controller in a virtual environment can interpret software commands the same way as the real controller of the system. The operator, who is trained in a virtual environment, can immediately take over the control of a robotic cell since he knows all its properties and behaviors.

Since the robot and the eccentric press are virtual replicas of real devices, such a simulation can also serve a technician during troubleshooting. In a virtual environment it is possible to show internal the structure of the system as well as disassembly and assembly procedures for individual parts.

Figure 10.6 shows the scene graph for a robotic cell. This graph is much more complex than the graph representing a game of table hockey. The graph also shows that the robot segments are connected in series (serial mechanism). A displacement of the first joint causes changes of positions of all robot segments. The eccentric press has even more components, so not all are shown in the scene graph.

During the design of complex virtual environments such as the presented robotic cell, it is necessary to determine the mutual relationship among all components of the system. So, for example, rotation of the eccentric press main motor affects the movement of gears and the entire tool section of the press. For proper functioning of the simulation, it is necessary to know the relationship between displacements, which are determined by the kinematic model of the device.

Fig. 10.7 Modeling of a mechanism with a skeleton: **a** initial block representing the trunk. **b** Extrusion of legs from the trunk. **c** robot model and **d** Skeleton that allows displacement of robot segments (t trunk, h head, a_r *right* arm a_l *left* arm, l_r *right* leg and l_l *left* leg).

10.3 Modeling and Simulation of an Avatar

In addition to various objects, virtual environments often contain avatars (representations of users in a virtual environment). We will next analyze modeling and simulation of movements of a simple avatar. Although graphical modeling was not specifically addressed so far, we will briefly explain the process since it is important for understanding of animated avatars.

The process of graphical modeling of a simple robot is shown in Fig. 10.7. The first step shown in Fig. 10.7a represents a block that will form the basis of the robot trunk. Legs are extracted from the trunk in the next step (Fig. 10.7b) and then the procedure is repeated for the hands and head (Fig. 10.7c). The final step is the most important. The robot is now modeled as a rigid body, which does not allow articulation in the joints. To enable animation of motion, it is necessary to generate a robot skeleton, which represents the joints and segments of the mechanism. For this simple example we added one joint and one segment in each limb. The trunk and head

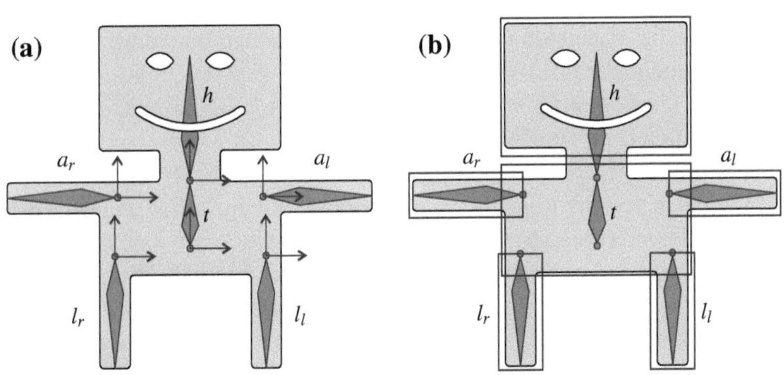

Fig. 10.8 Placements of coordinate frames (**a**) and bounding volumes OBB for collision detection between a robot and other objects (**b**) (t trunk, h head, a_r *right* arm a_l *left* arm, l_r *right* leg and l_l *left* leg)

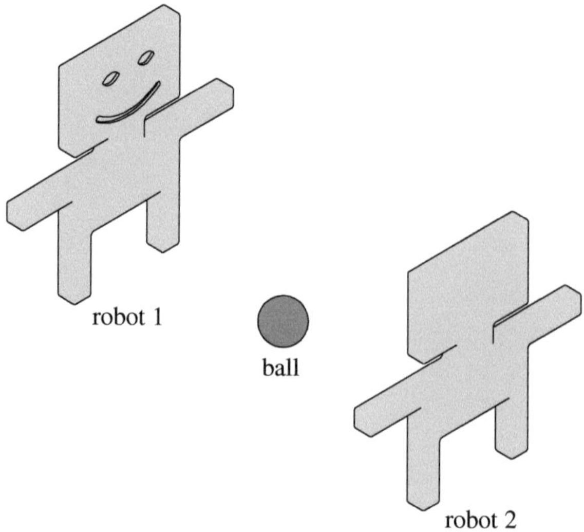

Fig. 10.9 Two robots and a ball

are also each represented with a single segment. There are altogether six segments that are connected by five joints. The central segment to which all other segments are attached is the trunk. The skeleton forms the basis for animation of motion. A coordinate frame is attached to each segment, and animation of avatar movement can be achieved through transformations of coordinate frames (Fig. 10.8a). If the surface of the robot continuously transforms (bends) across the robot joints, this generates an appearance of a skin covering the avatar.

Models of avatars are relatively complex because they usually contain a large number of degrees of freedom. Collision detection between avatars and surrounding objects thus also becomes computationally intensive. For this purpose, the model of an avatar can be simplified with the use of bounding volumes. In the simplest case, the entire avatar can be embedded into a single OBB (oriented bounding box) or AABB (axes aligned bounding box) volume. Figure 10.8b shows simplification of robot geometry with six bounding volumes covering the torso, the head and individual limbs. Such detailed representation of geometry allows a more accurate collision detection. At the same time, the computation remains relatively simple.

Figure 10.9 shows the concept of two robots playing a ball game. Since it is possible to move the robot segments and detect collisions between the ball and the robot segments, it is consequently possible to kick or throw the ball, thus allowing implementation of different games (football, volleyball, tennis).

Figure 10.10 shows the scene graph that includes both robots and the ball. The scene graph is relatively simple. The graph shows that the robot trunk is the basic structure and coordinate frame that determines position and orientation of the robot in space. Other segments are attached to the trunk.

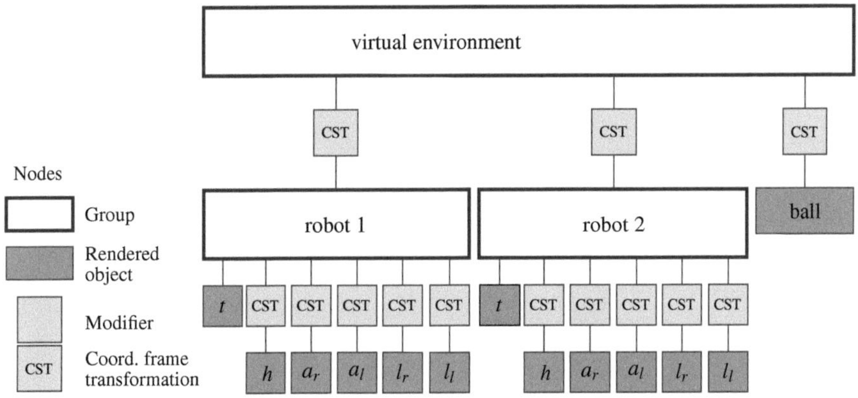

Fig. 10.10 Scene graph for two robots and a ball (*t* trunk, *h* head, a_r *right* arm a_l *left* arm, l_r *right* leg and l_l *left* leg)

10.4 Interactive Education Methods

Humans perceive the world in three dimensions since this is their natural environment. The perception of the world in three dimensions enables rapid learning and gaining of new experiences. Most education, simulation and training is still based on the presentations of text and two-dimensional images, where each person has to create his own visual representation. In doing so, not all people are equally successful. By using three-dimensional interactive technologies it is possible to replace the process of creating own representations by imposing correct visualizations. By requiring direct interaction with the presented content it is possible to further accelerate the learning process. This shortens the time required for education and training, improves quality and indirectly saves financial resources.

Pictures often tell more than a thousand words. Something similar could be said for virtual and augmented reality: interactive three-dimensional models tell more than a thousand pictures. An interactive model can be manipulated, its inside structure can be uncovered and its functional elements represented in their active roles. A model can be looked at from different angles and, most importantly, shows the actual functional operation of the object that it represents. These models are consequently very suitable for teaching since the world is represented in a natural way.

Figure 10.11 shows the spatial visualization of functional units of the human brain. With a simple manipulation it is possible to uncover and look at each functional unit. At the same time it is possible to get a description of each part and look at the brain from different perspectives. If the model is displayed on a system which allows spatial visualization, it is very easy to perceive and interpret spatial relations between different functional units.

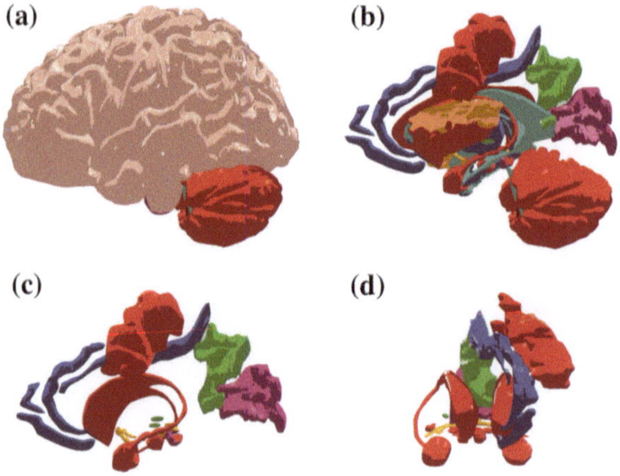

Fig. 10.11 Spatial representation of various functional units in human brain: **a** full brain model, **b** internal functional units, **c** selected internal functional units and **d** internal functional units from a different perspective

10.5 Interactive Configuration of Products

One of the most important fields of application of virtual reality is configuration of products according to user's requirements and preferences. This option is useful in those industries where possible configurations of products are varied and the cost of manufacturing demonstration products so high that it is not possible to produce all possible configurations among which customers could choose. Typical such areas include the aerospace industry (aircraft interior configuration), automotive industry (choice of materials, colors and accessories), architecture and furniture industry (choice of furniture configurations, colors, textures and functionalities of individual elements).

More and more manufacturers use large stereoscopic displays during design and evaluation of virtual prototypes and for demonstrations of their products to potential customers. Findings from preliminary studies are taken into account during the design of final versions of computer models. These can also be displayed in a virtual environment, where models come to life in three dimensions. This provides possibilities for verification of drafted concepts and for consultation regarding the next design steps.

Planning and design in the broadest sense includes engineering design of assembly components, products or buildings. When objects need to be optically and aesthetically evaluated, a three-dimensional visualization can provide valuable support. Modern technology provides designers, constructors and architects with the opportunity to more easily examine the functionality of their concepts and identify possible errors. New findings enable further optimization and a high degree of perfection of the product at an early stage of planning and design.

Communication between members of the development team is crucial. Three-dimensional visualization contributes to mutual understanding of constructors and designers, resulting in an improved workflow. Presentation of the final product using three-dimensional techniques can also persuade customers even before the actual prototype is built.

In the automotive industry, car producers are repeatedly faced with the challenge of reducing the time between the creation of a new design idea and its introduction into the market in the form of a new product. Software tools and hardware are essential in the planning, design and manufacturing of new models of cars. Designers and engineers often rely only on visual feedback during their work. That is why the choice of an appropriate system for realistic visualization has an important role in reducing the time required for bringing a product on the market.

Architects have long been using computer programs for the design and development of their products. Three-dimensional display technology now makes it possible to test new ideas very early in the process and prior to the actual construction. Such approach can be based on the use of stereoscopic displays, which cause the product to pop out of the drawing plane. Technology for displaying three-dimensional images

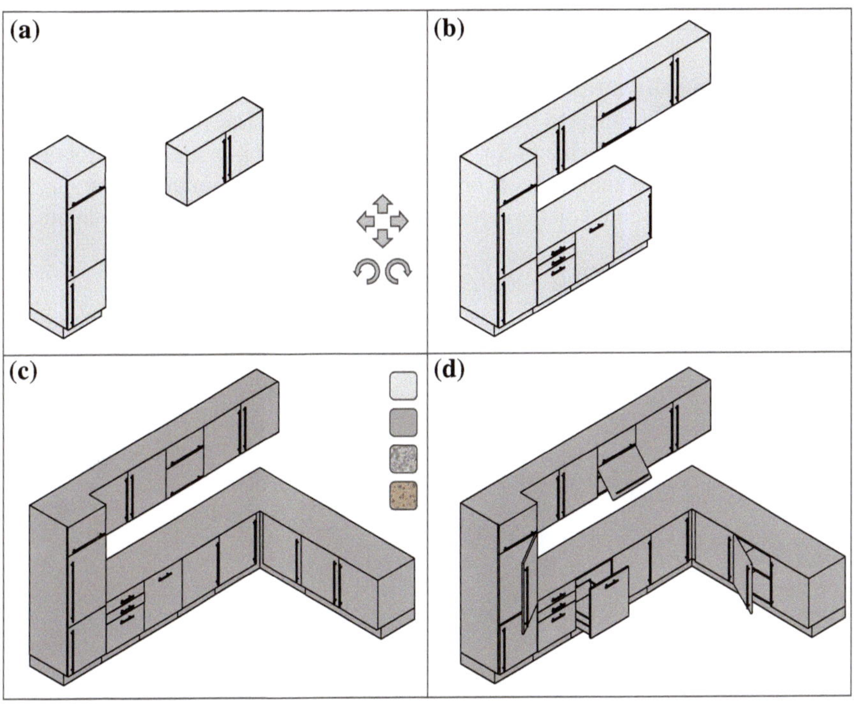

Fig. 10.12 Furniture configuration application: **a** initial assembly and positioning of an element, **b** intermediate assembly, **c** final assembly and selection of colors (textures) and **d** display of functionalities of various assembly elements

allows products to be presented in three dimensions without being forced to build costly and time-consuming prototypes. Changes in blueprints can thus be observed almost in real time.

Figure 10.12 shows the concept of configuring furniture elements. The user must be able to choose between different elements, must have the possibility of positioning elements in space, and should be allowed to select colors, textures and other parameters that define the functionality of the furniture assembly.

The result of the product configuration are the instructions for the manufacturer of the product that define the items that will be used in the manufacturing of the final product.

The concept of product configuration can be further extended, for example, to the case of the textile industry, where a three-dimensional model of the customer (which can be obtained by optical scanning) could serve as a basis for virtually trying on different clothes an shoes without the need to visit the store.

10.6 Medical Applications

In medicine, three-dimensional displays are increasingly used for training and diagnosis.

Modern minimally invasive surgery is often based on the use of a combination of endoscope, laparoscope and three-dimensional display. The surgery is performed through a small incision in the body. The display provides a spatial image obtained in real time through an endoscope or other medical imaging techniques. The surgeon uses information presented on the display to control the laparoscope either manually

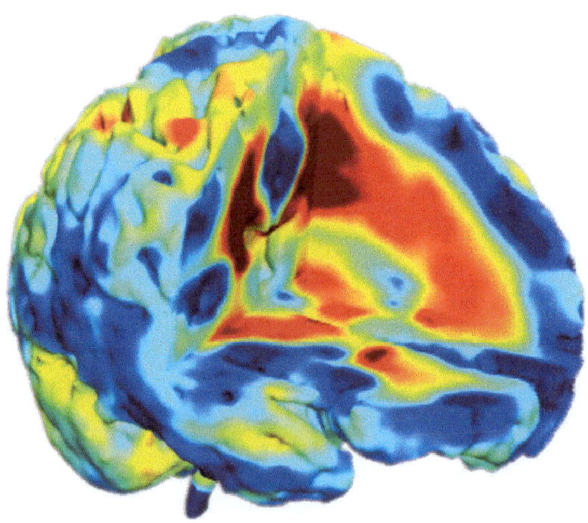

Fig. 10.13 Brain image from a nuclear magnetic resonance scanner

or through robotic assistance. Such assisted intervention provides a significant improvement in the ergonomics of the surgeon's workspace.

Data obtained by nuclear magnetic resonance imaging (MRI) and computerized tomography (CT) inherently include the third dimension (Fig. 10.13). Technologies that enable reproduction of three-dimensional images are therefore the most natural way for displaying such content. Diagnostics can be performed much more accurately and reliably using three-dimensional display of captured data. Such diagnostics can be used by doctors before, during or after the surgery.

Use of augmented reality allows superposition of information about the location of surgical instruments relative to a three dimensional image of a patient's body. The surgeon is thus always provided with a virtual view of the surgery relative to the target zone.

Index

M. Mihelj et al., *Virtual Reality Technology and Applications*,
Intelligent Systems, Control and Automation: Science and Engineering 68,
DOI: 10.1007/978-94-007-6910-6, © Springer Science+Business Media Dordrecht 2014

MIX
Papier aus verantwortungsvollen Quellen
Paper from responsible sources
FSC® C105338

If you have any concerns about our products,
you can contact us on
ProductSafety@springernature.com

In case Publisher is established outside the EU,
the EU authorized representative is:
Springer Nature Customer Service Center GmbH
Europaplatz 3, 69115 Heidelberg, Germany

Printed by Libri Plureos GmbH
in Hamburg, Germany